FLEXRAY AND ITS APPLICATIONS

FLEXRAY AND ITS APPLICATIONS
REAL TIME MULTIPLEXED NETWORK

Dominique Paret
dp-Consulting, Paris, France

Translated by Bill Chilcott
Fellow of the Institute of Translation and Interpreting
Claygate, Esher, UK

A John Wiley & Sons, Ltd., Publication

Contents

Preface

Why this book now?

Today, protocols for multiplexed industrial networks such as CAN, LIN and others are relatively mature, and only a few aspects such as 'Time-Triggered Protocol' and 'X-by-Wire' systems continue to evolve.

On the two latter subjects, little information or technical training is available to engineers, technicians or students. We hope that this book will at least partly fill this gap.

I waited for a long time before again dipping my pen into the inkwell of my PC(!). I preferred to wait until there were no announcements of the 'free shave tomorrow ...' type in sight. Which, of course, as usual, took a long time ... Version 2.1, revision A of FlexRay was delivered officially to the public in March 2005, then some minor modifications and additions (conformity tests) called rev. 2.1 A and B were added in November 2006, and finally, at the end of 2010, there was 3.0, which clarified some points of detail.

Above all, this book is not intended to give a literal translation of the standard, the original version of which can be downloaded free from the official website of the FlexRay Consortium (www.flexray.com). Instead, its aim is to act as an introduction and a detailed teaching presentation of the technical principles and operation of the FlexRay protocol. It is also aimed at giving newcomers an overall view of the concepts and applications.

The aims which FlexRay was intended to achieve (speed and security of communication, flexibility in operation, real time, distributed intelligence, network topologies, and so on) made it necessary to design the structure of the communication protocol so that it is directly related to the physical performance of the physical layer. When you read this book, always keep in mind the concerns generated by the physical layer (the medium and its management). Ideally, just as in music (see below), it would be necessary to present the communication protocol and the physical layer and their interactions simultaneously and in parallel ... which is mechanically difficult for a publisher, however experienced!

Something else you should know is that the content of the FlexRay protocol is dense, and includes numerous technical concepts which collide with each other, become confused with each other and intersect with each other, which makes it difficult to choose a plan for presenting the various chapters.

Author's note

To cover this subject of 'multiplexed networks' correctly, this book describes many patented technical principles which are subject to the operation of licences and their associated rights (bit coding, communication techniques, and so on), and which have already been published in official, professional technical texts or communications, or during public conferences or seminars – but above all, which must be used according to the legal rules in force.

How to read this book

In a previous book (*Multiplexed Networks for Embedded Systems: CAN, LIN, FlexRay, Safe-by-Wire*), we provided an overview, which was complete at the time, of this evolving field, using long technical introductions on these subjects. Today, this book, which is entirely about FlexRay, is dense because virtually all the 'real' subjects – principles, components, applications, security, and so on – are approached in practical terms. Also, to avoid discouraging the reader who is trying to understand these devices, we have put great stress on teaching so that the link between theoretical, technological, economic and so on aspects can be constantly established.

The challenge is therefore to present everything in the clearest, most suitable manner. After long reflection and long oscillations,[1] it has been necessary to choose a comprehensive presentation so that you, the Reader, can find your way easily through the maze of all these emerging communication principles and new protocols. We also advise you, before approaching the technical details which will be explained in the following chapters, to take the trouble to read the next few lines, which are intended to explain the why and how of the plan of this book and how to use it.

The aim of the introduction is to make your mouth water by giving a general survey of the applications which daily affect the motor vehicle and embedded systems of all types. Obviously, everything we have written in this book can be generalised to industrial applications of all kinds (control of machine tools and production lines, avionics, rail transport, building automation, transport of digital images, and so on).

The first part (A) is a reminder of the CAN protocol, a quick mention of the operation and contents of the TTCAN protocol and a review of the latest applications of 'X-by-Wire' type. We will briefly discuss the functional and application limits of CAN, and we will consider 'event-triggered' and 'time-triggered' communication systems, and all the implications which that consideration generates for so-called 'secure real time' applications.

In the second part (B) we will present, progressively, FlexRay and its protocol, in terms of communication cycles, decomposition of cycles into frames, format and content of frames, omitting any consideration of clock synchronisation between nodes.

Then, in the third part (C), we will go on to the analysis of the physical layer and the basic concepts of bit coding, propagation and topologies which can be used, and their effects. The problems of network synchronisation in operation and during the wakeup and startup phases are the subject of the fourth part (D). We will consider the architecture of a node, components of a FlexRay network, AUTOSAR and the range of associated hardware and software tools for providing support for development simulations, verification stages, production, maintenance, and so on in the fifth and final part (E).

---

[1] All $(1 - \Gamma^2)$ and (voltage) standing wave ratios included (*naturellement*).

A little music in this brutal world

Let us finish this introduction on a lighter (musical) note. Very serious discussions with some friends and FlexRay designers one day led us to the conclusion that a FlexRay system could be seen as a little like a musical score. The protocol description represents the melody in a treble key, and the physical layer represents the accompaniment in a bass key. In fact, with FlexRay as for reading a musical score, it is necessary to succeed in following the score not only by reading the two horizontal staves simultaneously, but also by reading the score 'vertically', to recreate the whole correctly. Additionally, like any well-informed musician, it is necessary to read ahead while playing! Welcome to FlexRay for musicians and future musicians!

I wish you good and productive reading throughout the pages of this book – above all, enjoy it, because I didn't write it for myself but for you! If there is still a shadow of a doubt about the subject and form of this book, your (constructive ☺) comments, remarks, questions and so on are always welcome by e-mail to my address, dp-consulting@orange.fr.

Thanks

The subject of multiplexed communication systems and networks is growing day by day, and very many skilled people are working in these fields. Luckily, I have had the opportunity to meet many of them, and consequently it is very difficult for me to thank everyone individually.

Nevertheless, I must address some special thanks to several groups of people:

- To numerous colleagues and friends of Philips/NXP Semiconductors of Nijmegen (Netherlands) and Hamburg (Germany), with whom I have had the pleasure of working for long years on these subjects, and, taking the risk of making some people jealous, more particularly Messrs Hannes Wolff, Bernd Elend, Thomas Shuermann, Peter Bürhing, Peter Hank, Burkhard Bauer, Karsten Penno, Patrick Heuts, Matthias Muth, the numerous 'Hans' and other colleagues in the Netherlands, and the numerous 'Peters' and other colleagues in Germany.
- To the long-standing international friends in the field of multiplexed buses in motor vehicles, Messrs Florian Hartwich, Bernd Müller, Thomas Führer of the R. Bosch company, Florian Bogenberger of Motorola/Freescale and Wolfhard Lawrenz of C & S.

It would be ungrateful not to thank also the numerous colleagues in the profession, and motor vehicle and equipment manufacturers, whom I meet regularly either at working meetings or at ISO, for their remarks and comments about the editing of this book, thanks to whom we all hope that this subject of multiplexed buses will blossom as it deserves.

Finally, I am very glad to thank Ms Manuela Philipsen of the 'Marcom' team of NXP Semiconductors in Eindhoven, for the numerous documents and photographs which she has been kind enough to supply to me for years to illustrate these books. Even more finally, I am also immensely grateful to the Vector Informatik GmbH company of

Stuttgart (Mr Uwe Kimmerley and the whole FlexRay team) and Vector France SAS of Paris (Mr Henri Belda, Mr Jean-Philippe Dehaene, Ms Hassina Rebaïne and Ms Rim Guernazi) for their technical and logistical support, their participation in the editing of certain chapters and for having had the kindness to agree to supply numerous very fine educational illustrations to enrich this book. In fact, this type and quality of teaching aid is fundamental to good distribution of knowledge, and in Vector's case is part of very rich support for professional training which is useful for spreading a technique and a technology. Setting up such support requires a large investment in time and money, and authorisation to publish them – even in part – really deserves to be welcomed as much as the quality of their content. So a big thank you for having done and authorised that.

Dominique Paret
dp-consulting@orange.fr

List of Abbreviations

ABS	anti braking system
ADC	analogue to digital conversion
AP	action point
Autosar	automotive open system architecture
BD	bus driver
BER	bit error rate
BSS	byte start sequence
CAN	controller area network
CAPL	Communication Access Programming Language
CAS	collision avoidance symbol
CC	communication controller
CDD	CANdela Data Diagnostic
CDF	cumulative distribution function
CHI	controller-host interface
CID	channel idle delimiter
CPU	central processing unit
CPU/ECU	central processing unit/electronic control unit
CRC	cyclic redundancy check
CSEV	channel status error vector
CSS	clock synchronization startup
DLC	data length coding
DLL	dynamic link library
DPI	direct power injection
DTS	dynamic trailing sequence
ECC	error correction code
ECU	electronic control unit
EMB	electromechanical braking
EMC	electromagnetic compatibility
EPL	electrical physical layer
EPLAN	electrical physical layer application notes
ESD	electrostatic discharge
FES	frame end sequence
FIFO	first in, first out

FSEV	frame status error vector
FSS	frame start sequence
FTDMA	flexible time division multiple access
GTDMA	global time division multiple access
HIL	hardware in the loop
LIN	local interconnect network
MTs	macroticks
MTS	media access test symbol
NIT	network idle time
NRZ	no return to zero
OEM	original equipment manufacturer
OS	operating systems
OSI	open systems interconnection
OSI/ISO	open systems interconnection/International Standard Organisation
PDF	probability density function
PDUs	protocol data units
PLL	phase locked loop
PS	protocol specification
RF	radio frequency
RTE	run time environment
SBCs	system basic chips
SD	dynamic segment
SHF	super high frequency
SOC	start of cycle
ST	static segment
SW	symbol window
SWCs	software components
TDMA	time division multiple access
TRP	time reference point
TSL	test set library
TSS	transmission start sequence
TTCAN	time-triggered communication on CAN
TT-E	time-triggered external
TT-L	time-triggered local
TTP/C	Time Triggered Protocol Class C
Tx	transmission
UHF	ultra high frequency
VDR	voltage-dependent resistors
VFB	virtual functional bus
WCET	worst case execution time
WUP	wakeup pattern
WUS	wakeup symbol

Part A

'Secure Real Time' Applications

1

Reminders about the CAN Protocol

As an introduction to this chapter, we will remind you of some general points about all the architectures of embedded systems, and starting from now we will take a very surprising turn by passing judgement on the properties of the well-known controller area network (CAN) protocol, presenting its principal limitations and finally imagining solutions which open up new horizons for decades to come.

1.1 The Limitations of CAN

Firstly, the concept of CAN, which was designed almost 30 years ago, is perfect for current applications, and will remain perfect for very many applications. However, time passes, and some of the inherent limitations of CAN, which have been known since its genesis, are now clearly visible. They are:

- **Limitations of bit rate** – Since it began, the maximum gross bit rate of CAN has been limited to 1 Mbit/s, and forthcoming and future application fields of embedded multiplexed networks require higher gross bit rates, of the order of 5–10 Mbit/s, either for purely functional reasons of timing or because of saturation of communication capacity. Everything must therefore be rebuilt. The word 'everything' in the previous sentence probably surprises you, but it's true! In fact, everything must be rethought and rebuilt, because 1 Mbit/s, the maximum bit rate value for CAN, corresponds in practice to the limit of a technical philosophy in which it was still possible to avoid talking too much about the phenomena and/or effect of line propagation, reflection coefficient, stubs, Smith's abacus, and so on. Beyond this value, when designing protocols and their physical layers, it is impossible to avoid considering and taking account of these physical parameters and their effects.
- **Limitations of distance and topological flexibility** – It should also be noted that the 1 Mbit/s maximum value of CAN is related to the structure of the acknowledgement bit of the protocol. In fact, so that the protocol functions correctly, it is necessary to be certain that the sum of the outgoing and incoming times of the signal allows

FlexRay and its Applications: Real Time Multiplexed Network, First Edition. Dominique Paret.
© 2012 John Wiley & Sons, Ltd. Published 2012 by John Wiley & Sons, Ltd.

the acknowledgement signal to fall within the duration of the bit. This special feature of the protocol imposes limits on the propagation time, and therefore a maximum distance, between nodes which are present on the network, but it also excludes some topological possibilities and solutions involving propagation asymmetries according to which branches of networks are used.

- **Limitations of the possibility of topological redundancy** – This point is linked to the two previous ones (distance and acknowledgement). Creating a network which makes it possible to provide redundancy of communication at the level of physical layers according to a CAN architecture/topology is difficult, not to say impossible. Consequently, it seems futile to hope to implement systems which are entirely controlled using links which function according to the famous 'X-by-Wire' (everything by wire) concept, which we will describe in detail in a later chapter.
- **Limitations of access to the medium in real time** – As we will show and/or remind you later, CAN has a strong 'event-oriented' orientation. The phases of communication on the network are mainly initiated by sporadic, random, probable, and so on events. Also, CAN lacks a 'real time' orientation, or in other words a philosophy with a 'time-oriented' orientation; that is, one in which the communication phases are initiated as a function of a clock, a date, a fixed instant. To get round that while preserving the structure of CAN, one of the first responses was the creation, by the R. Bosch company, of a higher-level application layer called 'TTCAN' or time-triggered communication on CAN, which is initiated by events in time, to refresh CAN a little (see Chapter 2).

It should be noted that all these points have been covered by the appearance of the FlexRay concept, which we will describe in detail later.

1.2   'Event-Triggered' and 'Time-Triggered' Aspects

1.2.1   The Probabilistic Side of CAN

By its design and structure, the CAN protocol encourages transmission of communication frames when events occur at a node of the network. This is what is called an 'event-triggered' system. In fact, often a participant sends a message following an action, a reaction or a request for information as a function of the requirements of the intended application and/or of its own task.

As we explained in numerous previous works, CAN messages are prioritised (offline) by the system designer, using values which the designer has chosen to assign to the identifiers of the communication frames. On this principle, at a given instant, no node can be certain that its message is transmitted immediately, because of the conflict management and arbitration resulting from the values of the competing identifiers at this precise moment of access to the network. This type of concept and the management of it give transmission of messages on the network in CAN a strong 'probabilistic' emphasis, because it is subject to the arbitration procedure. The latter is a function of the respective values of the competing message identifiers at the time of the attempt to access, and then seize, the bus or medium, which makes the timing of this transmission – and the associated latency time – very dependent on the probability of the appearance of the respective values of the identifiers.

The only true CAN message which is truly 'deterministic' is the message with the identifier hex 0000, since, for this identifier value only, the latency time is strictly known, and its value is '*one CAN frame minus one bit plus the inter-frame time (3 bits) ...*', since, to within a bit, this (highest priority) message could not access the network last time round.

For other messages (identifiers other than hex 0000), that depends on big ideas of scheduling, obscure calculations of probability applied to the respective values of the activity model of the network, and to the appearance of the respective values of the competing message identifiers. Also, the probability of this arbitration phase taking place is excessively high, since each time the medium is occupied – as is very often the case – all the other nodes which have been unable to access it wait for the propitious moment to try to get it back, and all starting at the same moment, just after the inter-frame phase required by the CAN protocol, are all immediately subjected to the arbitration procedure.[1]

The problem then occurs when what is wanted is to communicate – transmit or receive – definitely, at a precise, predetermined instant, so that the timing is deterministic. In principle, nothing in CAN permits this. Consequently, it is necessary to create new systems, certain actions of which are triggered spontaneously at precise instants. These are usually called 'time-triggered' (TT) systems; that is, in our case three principal concepts, TTCAN, TTP/C (or Time Triggered Protocol Class C) and FlexRay, which we will explain in detail below.

1.2.2   The Deterministic Side of Applications

In very many applications, it is or becomes necessary to trigger certain actions spontaneously at precise instants. Such systems are called 'time-triggered' or systems functioning in so-called 'real time' mode. When systems must function in 'real time' (which in principle does not exist and is merely an abuse of language[2]), the big problem occurs when what is wanted is to be sure of communicating – transmitting or receiving – at a predetermined instant, or in specific time slots, thus adding a 'deterministic' aspect to communication.

As already mentioned, in principle nothing in CAN makes it possible to guarantee this. In these cases, it is therefore necessary to set up a mini real time 'operating system' of TT type, for example on the higher layers of the OSI (Open Systems Interconnection) model (at layer 5, 'session' or layer 7, 'application') or to integrate or encapsulate this type of function into a definition of the protocol which is capable of solving all or part of this problem.

To do this, customarily, so that information can circulate on the network, specific, well-defined time windows are used. How these time windows are implemented is, in principle, completely free and non-limiting. The only specific point consists of ensuring that all the participants are perfectly synchronised, so that each can talk or respond in its turn without overlapping into the time windows of its neighbours. To do this, it is generally necessary either to transmit a 'reference clock' cyclically to the whole network so that each participant resets its clock or to synchronise the clocks of all participants.

---

[1] We refer anyone who is interested in this subject to the numerous publications written by Mr Laurent George.

[2] To remove any doubt, the term 'real time' is customarily understood as actually implying 'time with known latency', that is it means that one is certain that at a precise instant the thing which is supposedly being done actually is. Also, very often the terms 'real time' and ideas of 'deterministic' systems are confused. How, in certain deterministic conditions, the whole functioning can be assimilated to an idea of 'functioning in quasi real time' – that is, with deterministic access to the communication medium and known latency times – will be explained below.

2

The TTCAN Protocol

In the early 1990s, the dominant position of CAN in the market, and the increasing complexity of embedded systems, rapidly caused a demand for protocols which guarantee responses in 'real time', deterministic solutions and greater security. Consequently, systems using 'Global Time' devices were designed. The first of them which was actually used in industry, in the automotive field, was the 'TTCAN' (time-triggered communication on CAN) protocol, which was proposed by the R. Bosch company and the 'CAN in Automation' (CiA) group in TC 22/SC 3/WG 1/TF 6[1] of ISO, before becoming, in early 2002, ISO Standard 11898-4.

2.1 TTCAN – ISO 11898-4

TTCAN forms a protocol layer at a higher level than that of CAN itself, without in any way modifying the structure of the data link layer (DLL) and physical layer (PL) of the latter. To do this, TTCAN is placed mainly at the level of layer 5, called 'session', of the Open Systems Interconnection/International Standard Organisation (OSI/ISO) model, which in other words comes back to saying that the structure of the TTCAN protocol was designed to be encapsulated in the transport protocol of CAN.

The aim of TTCAN is to define and guarantee the latency time of every message at a specified value which is independent of the load on the CAN network itself. This protocol can be implemented at two levels:

- level 1, which is limited to transferring cyclical messages only;
- level 2, which supports a system called 'Global Time'.

Given that the aim of this book is not to describe this particular standard in detail, we refer you to the official documents for fuller information. However, we will summarise the broad outline in a few paragraphs.

It should be noted that TTCAN comes between CAN and FlexRay, and that use of it can make it possible – in certain applications – to reduce the load (over time) on some

---

[1] For those who do not speak ISO fluently, this means 'Technical Committee 22 (Road vehicles), Subcommittee 3 (Electrical and electronic equipment), Working Group 1, Task Force 6'.

FlexRay and its Applications: Real Time Multiplexed Network, First Edition. Dominique Paret.
© 2012 John Wiley & Sons, Ltd. Published 2012 by John Wiley & Sons, Ltd.

existing CAN networks and structures, or to regulate them. Its description corresponds to a session layer (number 5) of the OSI model (between layer 2, 'data link', and layer 7, 'application'), and is inserted into the original, overwritten CAN model. In short, to understand the CAN concept better, let us remind ourselves of the foundations of the session layer of the OSI model.

2.2   Session Layer

As a brief reminder, the OSI/ISO definition indicates clearly that: *'The purpose of the Session Layer is to provide the means necessary for cooperating presentation-entities to organize and to synchronize their dialogue and to manage their data exchange. To do this, the Session Layer provides services to establish a session-connection between two presentation-entities, [and] to support orderly data exchange interactions'*. This layer:

- on the one hand, carries out the necessary functions to support dialogue between processes, such as initialisation, synchronisation and termination of the dialogue, and so on;
- on the other hand, makes the constraints and characteristics of the implementations in the lower layers transparent to the user.

Thanks to it, references to different systems are made by symbolic names and not by network addresses. Also, it includes elementary synchronisation services and recovery at the time of an exchange.

2.3   Principle of Operation of TTCAN

TTCAN is based on a timed deterministic exchange, which is itself based on pre-established time windows of an operational cycle, also pre-established, and the overall operation of which can be visualised with the help of the matrix of lines and columns shown in Figure 2.1. This matrix summarises the general principle of the operation of this protocol.

All messages that must circulate on the network between the CAN nodes are organised like elements of an X by Y matrix. This system in the form of a timing matrix consists of time windows which are organised in 'basic cycles', with identical time values (shown by the whole of each of the lines X of the matrix), and in numerous time zones (windows) during which transmission is authorised (shown by the columns Y of the matrix). This matrix system thus defines the correlation between the time windows and the presence of messages on the network.

The principle of operation of TTCAN assumes that one of the nodes of the network is responsible for the organisation of the slicing and the time assignments which are considered. In fact, when the system starts, this node assigns to every node the time phase(s) which are reserved for it.

Consequently, the system becomes deterministic, since each of the nodes has the right to express itself at a precise moment, which it knows, and for a well-determined length of time. Obviously, this does not at all constitute a real time system, but if all the cycles

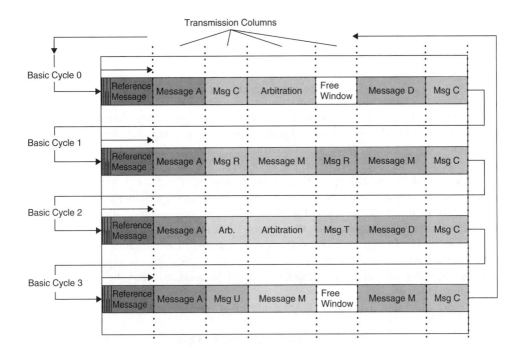

Figure 2.1   General principle of operation of the TTCAN protocol

are covered in full sufficiently quickly, the system quickly comes back to the same node, and this appears to all the participants like access to the network in 'quasi real time'.

To be more explicit, here is an example, greatly exaggerated but quite representative of the principle which is used:

- Being the system manager, I call myself node number 1, and I decide unilaterally to assign the following time phases to the four other participants in the network. To do this, I begin by communicating to them a minimum of necessary information for the whole to run well, using a generic message called 'reference message' which uses a specific identifier and indicates that:

   "From now until further notice, the duration of the basic cycle will be 1 hour, and here is the time window sequence for each of you:

   you, node no. 2, you don't have much to say, you will talk from the hour (hh.00) to hh.05;

   I, node no. 1, since everyone knows I'm a chatterbox, I'll talk from hh.05 to hh.20;

   you, node no. 3, you'll talk from hh.20 to hh.25;

   if you want to, everyone can talk from hh.25 to hh.30, under arbitration by CAN;

   you, node no. 2, you still don't have much to say, you can talk again from hh.30 to hh.35;

   you, node no. 4, you'll talk from hh.35 to hh.45;

you, node no. 2, you can talk again from hh.45 to hh.50;
you, node no. 5, you can talk from hh.50 to hh.55;
from hh.55 to the hour, nothing definite – everyone welcome – under arbitration
    by CAN;
and so that you all set your watches, I inform you that it is now very exactly 10.54."

Through this very embellished example, we hope that you have understood the basic mechanism of TTCAN. You have certainly noticed in passing that, to avoid harmful effects caused by drift of the clock of each of the participants, each basic cycle begins with a reference message. Also, within one cycle, we have allowed node number 2 to speak several times, although it has little to say each time, but it has to provide information frequently. How time is distributed is, in principle, absolutely free, and left to the goodwill of the cycle manager. For obvious reasons of synchronisation and possible drift, the 'time master' must send the reference message periodically. Also, you will notice that the electronics of each node must be capable of maintaining a certain clock precision throughout the duration of a cycle, so that it does not overlap with the other participants, since new clock information will not be received until the start of a new communication cycle.

In a more structured manner, TTCAN defines that (see Figure 2.1):

- periodic messages are included in 'exclusive time windows';
- sporadic messages are included in 'arbitration time windows';
- 'free time windows' are reserved as spaces which are free of any movement.

Obviously, the daily reality in application is quite different, on the one hand because of a host of constraints because of the systems in use, and on the other hand because it is frequently necessary to reconfigure the time sequence because of external events, foreseen and not foreseen.

So there, summarised as simply as possible, in a few paragraphs, is the philosophy of the TTCAN concept. If you want more detail on this protocol, refer to:

- ISO Standard 11898-4 for the strictly factual side;
- CAN in Automation – CiA (www.can-cia.org), which can provide you with basic application support for use of TTCAN.

3

Emergence of 'X-by-Wire' Systems

What strange, barbaric words! 'High throughput': no problem! 'Redundant' and 'redundancy' indicate that some functions, message transmissions, physical media, and so on are doubled, tripled, x-times-ed to provide the desired security in operation. That leaves 'X-by-Wire', so let's look at that.

3.1 High Throughput and X-by-Wire

To satisfy the more and more greedy requirements for rapid processing of information, future systems and concepts must be capable of supporting high communication bit rates, with everything that implies in terms of performance of the physical layer, transmission quality, synchronisation between nodes, and so on.

The generic term 'X-by-Wire' includes all types of application which implement 'systems controlled by wire links', which are also understood not to have 'any other control which is carried out via a mechanical interface'. Actually this is not new! For several decades, numerous embedded systems used in the aviation industry have operated according to ' . . . by wire' models (for example the first Airbuses). For a long time, the control surfaces and flaps of aircraft have not been controlled mechanically using rods, hydropneumatic jacks and other mechanical systems. These controls have been replaced by electric motors which are controlled using wired networks, connected to each other and arranged in a bus topology or otherwise – and it even works correctly – otherwise, we would know! Surprising as it may seem, to ensure that the systems concerned are very safe, that can be done by taking a mass of structural precautions, sometimes without any physical and/or software redundancy.

3.2 Redundancy

Of course, to ensure a higher level of operational safety in all these systems, it is sometimes worthwhile to imagine devices with some supplementary redundancy, whether in communication, or in the physical layers, or in the two together. And in fact, the automotive world

FlexRay and its Applications: Real Time Multiplexed Network, First Edition. Dominique Paret.
© 2012 John Wiley & Sons, Ltd. Published 2012 by John Wiley & Sons, Ltd.

and other industrial application sectors are very interested in these technologies – with the start of mass production forecast for around the years 2012/2015 – for the same reasons as those which guided their predecessors in aviation, progressively replacing mechanical controls with controls 'by wire'. To begin with, goodbye to suspension springs that suffer fatigue (even on board a vehicle stopped on the roadside!), anti-roll bars to assist road-holding, steering columns that can pierce the stomach in the case of impact, steering racks, master brake cylinders that can leak, accelerator control cables that stick or break! And long live the weight reductions of equipment in vehicles – and therefore their

Figure 3.1

Figure 3.2

Figure 3.3

excessive consumption and pollution – and the improvements of passive security in case
of collision. And then, for example this technology will offer greater flexibility for overall
mechanical design and exterior and interior design of vehicles (see some futuristic photographs in Figures 3.1–3.3). It will even be possible, on a single model, to make minor
modifications so that vehicles with right-hand or left-hand drive are easily available, or
a more innovative design of the instrument panel, or the possibility of getting rid of the
brake and clutch pedals ... and finally, the constant reduction of cost. In short, the future
is ours!

After the fantastic predictions of our favourite crystal ball, which will become reality
in volume within 6–10 years from now, it is now necessary to think about action, as we
reflect on the numerous problems to be solved. That is what we propose to do now.

3.3 High-Level Application Requirements

Let us take up the story and examine our future embedded system. Let us begin by
observing the trends of tomorrow's vehicle architectures and embedded systems.

3.3.1 The Number of Communication Systems is Growing

In case you didn't know, a high-range car already (in 2011) counts between 60 and
75 central processing units (CPUs) (!), and also has five or six CAN networks (high
speed and fault-tolerant low speed) and six or seven local interconnect networks (LINs).
Consequently:

- the number of gateways between systems and networks is growing;
- the topologies of networks are more and more complex;
- there are more and more very high level interactions between the various systems.

3.3.2 The Electronic Architecture Must be Common to Several
Vehicle Platforms

The electronic architecture must be common to several vehicle platforms, so that it can
induce large synergies, rapid migration taking innovation with it and cost reduction. Every

motor vehicle and equipment manufacturer is becoming more and more specialised in its particular fields of skill, and the direct consequence is the need to be able to design the electronic and electrical architecture of the intended system in a manner which is modular and capable of evolving at a variable scale or geometry (the famous idea of 'scalable'). This structural elasticity (scalability) has implications at different levels:

- different brands and models of electronic modules must function on different platforms (effect on scalability and cost);
- creation of totally open interfaces (increase of the number of applications);
- 'application agreement' from end to end of product design;
- reduction of the complexity of systems by better-defined interactions between applications.

3.3.3   Some Things the Architecture of the Communication Network and the Nodes Must Allow

The architecture of the communication network and the nodes must allow:

- manufacture of low-range to high-range vehicles on the same platform;
- communication supported by single, dual or mixed physical communication channels;
- clear visibility of the network for the various application fields (chassis, safety, engine, environmental detection, and so on);
- use of inexpensive components (for example piezo-electric rather than quartz resonators).

Let us look quickly at the functional requirements of these new technical and industrial strategies.

3.4   High-Level Functional Requirements

From the start of the project, depending on what applications are intended, all the possibilities of optimising devices must be used, up to the physical limits of the principles that are used and of the components of the physical layer.

3.4.1   Speed of Communication

The quantity of information to be transported is much greater, and it is richer in content and quality. Consequently, the communication bit rate (1 Mbit/s) of the CAN high speed network which has been used until now is no longer enough. The gross throughput required for these new systems is of the order of 10 Mbit/s on a single-channel medium (with a net throughput of about 7.5 Mbit/s compared with 500 kbit/s for CAN) or on a dual-channel structure with a higher throughput and providing possible redundancy.

3.4.2   Physical Layer

The physical medium which is used for communication must be capable of being supported by at least two different technologies, for example one of wired type (e.g. differential pairs) and the other of opto-electronic type (e.g. plastic optical fibres), and must make it possible to put the network nodes into sleep mode, and wake them up, via the medium. Also, the signals that circulate on the physical layer must not pollute the radio frequency band (low emission of radio-frequency disturbance), and must be highly immune to interfering external signals. Additionally, 'containment errors' must be managed using an independent bus monitoring element, for example a physical or software 'bus guardian'.

3.4.3   Access to and Management of the Medium

From experience, in this type of concept, access to the medium in time is always an important and delicate point. To satisfy all the functional and security aspects of the network, the following are necessary:

- transmission of data of so-called 'static' or 'real time' type must be deterministic, for example using the principle of 'time slots' (e.g. time windows like those used in TTCAN);
- transmission of data of 'dynamic' type, triggered by events, must also be provided, to offer greater flexibility in use;
- there must not be any case of interference or interaction between the two transmission modes 'static' and 'dynamic' mentioned above;
- the chosen transmission principle must be completely free of any arbitration system (there can and indeed should be a prioritisation system, but don't confuse arbitration with prioritisation – which we will later call 'fighting' and 'fairness');
- the bandwidth (the bit rate) of the network must be adjustable, and it must be possible to allocate it dynamically;
- it must be possible to send different and/or complementary ('differential') information during the same time slot on two physically different communication channels;
- different nodes must be able to use the same time slot on different transmission channels.

3.4.4   Synchronisation Method

A reliable method of synchronisation – more precisely, globalisation of time – must be set up, to ensure that the operation of the various elements of the network is perfectly synchronised. To do that, several elements must be available, in particular:

- a device called 'Global Time', which carries out distributed synchronisation, triggered or not by a (physical, conventional or artificial) time reference ('time-triggered');
- synchronisation carried out by all the participants, helped by a master ('master synchronisation').

Also, the system must be capable of supporting momentary disappearances and reintegrations of nodes of the network, and cold and hot restarts of the network.

3.4.5   Network Topologies

As will be explained in more detail in Part C, if an increased bit rate and reliability of communication are wanted, the topological aspect of the network becomes more and more important! It is therefore necessary to consider using new topologies, other than the everlasting 'bus' configuration which has kept thousands of users alive until now. We will therefore consider topologies with:

- passive buses – just like before!
- passive stars;
- active stars, possibly put into cascades;
- a nice mixture of active stars and passive buses.

A fine syllabus, don't you think?

3.4.6   Requirements at System Level

On the basis of these different topologies and different performances which are required from the physical layer as explained above, it is also necessary to design a fault-tolerant system; that is, one which can tolerate faults and incidents in operation, has a dual transmission channel, detects its own errors and sends diagnostic messages. It is also necessary to think hard about setting up redundancies between the CPUs in each node, to ensure reliability of operation by multiple physical redundancies on the one hand, and redundancy of transmission on the other hand. Also, if only to make maintaining them, revising them, and so on easier, the systems must, sooner or later, be standardised (internationally by ISO if possible), interoperable, reusable (reuse being in fashion, as everyone knows), open to everyone, with no exclusive rights clause or payment of royalties, certified by recognised testing laboratories. In particular, they must offer a wide range of development tools during the design phase (emulators, simulators, and so on) and system integration phase (monitoring, fault injection, and so on), and of course have numerous component suppliers. In short, everyday stuff, the same old story – not forgetting the classic 'better for less cost'.

So we have quickly described the functional framework of these new networks, which, as you have no doubt noticed, are very distant from CAN but complementary to it. Let us now look at what proposals can respond to it.

Part B

The FlexRay Concept and its Communication Protocol

For teaching reasons (so that everything is not mixed up!), the purpose of this second part is to present only the most general aspects of the FlexRay concept. Parts C and D of this book will fill in the missing points of the concept in detail. This part therefore presents:

- in Chapter 4, the genesis of FlexRay;
- in Chapter 5, FlexRay and real time;
- in Chapter 6, the communication protocol;
- in Chapter 7, the modes and techniques for access to the medium.

A very specific application example offers a synthesis of the content of the chapters listed above. Also, as you will very quickly appreciate, there are many things that are not said and hidden techniques in one of the parts about modes of access to the medium. Which one? Wait and see! The technical appendix which resolves most of this unbearable suspense forms a whole separate chapter.

4

The Genesis of FlexRay

Before presenting the FlexRay concept and its genesis, let us mention, as a reminder, a solution which, for a few years, was claimed to be the solution to the problems mentioned in Chapter 3 for the automotive market. This is the Time Triggered Protocol Class C (TTP/C).

4.1 The TTP/C Protocol

The TTP/C system is one of the members of the large family of 'time-triggered protocols' (the '/C' indicates that it meets the criteria of Class C of the Society of Automotive Engineers (SAE) for the real time communication and fault-tolerance aspects of the automotive field). It was designed and developed by Prof. Hermann Kopetz of the University of Technology of Vienna, Austria, and was then taken over by the TTTech company[1] (plus some affiliates).

In short, TTP/C was designed on the principle that the strategy for access to the medium would be of the time division multiple access (TDMA) type, to which we will return in detail when we present FlexRay. This principle makes it possible to solve problems of interoperability between CPUs which are developed independently of each other. It should be noted that TTP/C was not originally dedicated to automotive applications, but was aimed at industrial applications generally.

On a certain date in 1997 or 1998, following some presentations which were made to the automotive world, for lack of anything better, some automotive manufacturers such as Audi (and Volkswagen, which is part of the same group) formed a 'TTA Group', where the A stood for 'architecture'. Other manufacturers such as BMW and DaimlerChrysler also worked on the TTP/C design for a few years, but left the group because they judged, for precise technical reasons, that the concept was not aimed sufficiently at the automotive field, and that working with TTTech was quite awkward.

After this interlude, let us now go on to examine FlexRay.

---

[1] Time-triggered technology (TTTech) Computertechnik GmbH is a system house which was formed to follow up the work of the Vienna University of Technology on the TTP/C communication protocol, to trade in it, and to provide a relationship between users and this technology for different applications.

FlexRay and its Applications: Real Time Multiplexed Network, First Edition. Dominique Paret.
© 2012 John Wiley & Sons, Ltd. Published 2012 by John Wiley & Sons, Ltd.

4.2   FlexRay

4.2.1   The Genesis of FlexRay

The genesis of FlexRay began with the formation of a group of industrialists who had decided to carry out an exhaustive technical analysis of existing networks which were used or could be used specifically in the automotive environment – that is, CAN, TTCAN, TCN, TTP/C, Byteflight (a proprietary protocol of BMW) – and to judge whether one of them was capable of meeting, for decades to come, all the technical and application wishes in the preceding chapters. The conclusions of this study clearly indicated that this was not the case, and this led to the definition of a new proposal, which was later called 'FlexRay'. In fact, in a few words, the inventory of existing solutions showed that:

- **CAN**
  - does not have a high enough bit rate for the new applications; it is difficult to make transmission really deterministic and redundant;
  - will not be replaced by FlexRay, but will work as a complement to it.
- **TTCAN**
  - in principle has the same throughput limitations as CAN;
  - regrettably is lacking support for optical transmission, a redundant transmission channel, 'fault-tolerant Global Time' and a bus guardian.
- **TTP/C**
  - has a frame whose information content is judged to be too low;
  - has few common properties with FlexRay, despite the use of bus access of the TDMA type;
  - does not offer or provide any flexibility relative to FlexRay concerning: the combination of the synchronous and asynchronous parts of transmission; the multiple slots for sending by the same node in the synchronous part; nodes acting on single, dual or mixed channels; a 'never give up' strategy regarding control of the communication system in relation to the application; the problems of the status of FlexRay members and of rights to licences and services.
- **Byteflight**
  - has too few functions;
  - can be functionally compatible with FlexRay if the latter is used in pure asynchronous mode.

4.3   The FlexRay Consortium

Following the technical report described above, in 2000, a 'Consortium Agreement' was signed between the following industrialists (look out for the first person who says 'oh hell, always the same people!'):

- the automotive manufacturers BMW AG, DaimlerChrysler AG and General Motors Corporation;
- the equipment manufacturer Robert Bosch GmbH;

- the silicon founders Motorola GmbH (which has since become FreeScale), which is mainly involved in the definition of the protocol and of the communication controller, and Philips GmbH (which has since become NXP Semiconductors), for the definition and the components of the physical layer, but also involved in the definition of the protocol;
- they were joined a little later by the manufacturer Volkswagen AG, to form the core partners of the FlexRay Consortium, each with the mission of bringing its specific skills; on the same occasion, the statuses of premium members and associated members were also defined.

THE GUILTY

When it's friendly, a little denunciation never hurt anyone! So here is the list of companies and people who worked together in the FlexRay Consortium so that the concept became a reality. As you will notice – and it is rare enough to be emphasised – to show the unwavering desire of the Consortium members to make progress, the (main) patents listed below were filed jointly by all the companies which are core members of the Consortium.

Publication date: 22 October 2003

Patent Numbers and Inventions

- EP1355458 Method for transmitting data within a communication system.
- EP1355461 Method and unit for bit stream decoding.

Applicants

- Bayerische Motoren Werke AG (Germany)
- Robert Bosch GmbH (Germany)
- DaimlerChrysler AG (Germany)
- Gen Motors Corp (US)
- Koninkl Philips Electronics NV (Netherlands)
- Motorola Inc. (US)
- Philips Intellectual Property (Germany).

Inventors

- Ralf Belschner, Josef Berwanger, Florian Bogenberger, Harald Eisele, Bernd Elend, Thomas Fuehrer, Florian Hartwich, Bernd Hedenetz, Robert Hugel, Matthias Kuehlewein, Peter Lohrmann, Bernd Mueller, Mathias Rausch, Christopher Temple, Joern Ungermann, Thomas Wuertz, Manfred Zinke (Germany)
- Thomas Forest, Arnold Millsap (US)
- Patrick Heuts (Netherlands).

Congratulations again to the parents of this beautiful baby!

For more information, refer to http://swpat.ffii.org/pikta/txt/ep/1355/461/

Figure 4.1 lists the various participants on a given date, and the levels at which they act. It should be noted that, as of the end of 2004, all major automotive manufacturers and players are in the FlexRay Consortium.

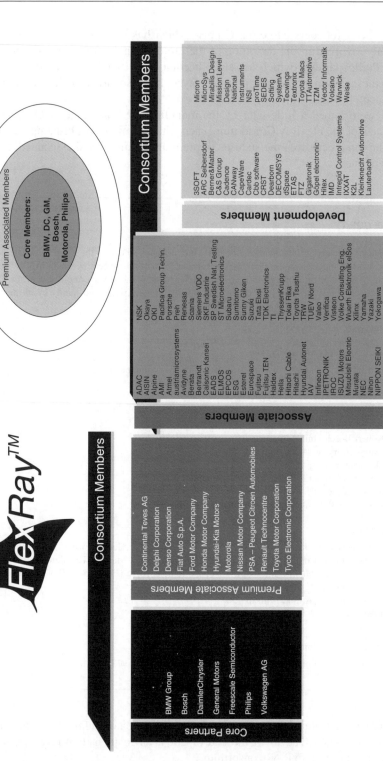

Figure 4.1   History of the creation of the FlexRay Consortium

Thus, BMW, DaimlerChrysler, General Motors and Volkswagen AG obtained the status of 'core members'. The status of 'premium associated members' was assigned to the following automotive manufacturers, in alphabetical order: Fiat, Ford, Honda, Hyundai-Kia Motors, Mazda, Nissan, PSA-Peugeot Citroen, Renault, Toyota, and so on, and to the equipment manufacturers Continental, Delphi, Denso and Tyco. The Consortium also includes at least fifty 'associated members'. Obviously, this means that the system has to be taken seriously!

It should be noted that the development tools company DeComSys (in which many people came from the team that developed TTP/C) was, for a long time, the official administrator of the Consortium. FlexRay was officially launched at a public conference in April 2002, in Munich, and the first 'FlexRay product day' took place in September 2004, in Böblingen. About 250 people took part in each of these events.

4.4 The Aim of FlexRay

4.4.1 A Flex(ible) Configuration

'The aim of the FlexRay Consortium is to create a communication system which is capable of controlling applications at four different levels:

- at high throughput for digital transmission; that is, being able to improve, complete and/or supplement applications which are limited by the bit rate of CAN;
- capable of implementing solutions of the 'X-by-Wire' type;
- being able to conceive solutions which provide redundancy:
  - thanks to the high throughput, by sending the same messages several times;
  - or by having available two distinct communication channels, transmitting the same information in parallel;
  - or by having two communication channels, which at a normal time transmit complementary information, in such a way as to have a throughput which is apparently greater than the physical bit rate of the protocol, and having as a fallback position full use of the remaining channel.
- and finally, capable of carrying out all future electronic functions in motor vehicles.'

In short, having available a flex(ible) configuration, which is the origin of the name 'FlexRay'.

4.4.2 Solutions

Let us now make a short inventory of what a FlexRay solution must be capable of resolving and providing, while trying to classify everything according to some large, more technical themes, of communication, topologies, security and applications (see Figure 4.2).

4.4.2.1 In Terms of Communication

FlexRay must:

- have a high throughput for transmission of digital data (10 Mbit/s);
- transmit data synchronously and asynchronously;

- cause to circulate on the network deterministic data transmissions, the latency times and jitter of the messages of which are known and guaranteed;
- have different simultaneous data throughputs, implemented using easy allocations of the pass band for each node of the network;
- be able to carry out static and dynamic segmentation of the data transmissions:
  - by distribution of requirements,
  - by distribution of functional domains.
- support:
  - a configurable number of time slots for sending per node and per operating cycle,
  - hardware and software redundancy with variable geometry (single and dual channel, mixed system),
  - putting the participants into sleep mode via the network, and waking them up,
  - management of the consumption of the participants in the network.
- detect and signal errors very quickly;
- support synchronisation faults/errors of the Global Time base;
- provide, by hardware devices, 'fault-tolerant' and 'time trigger' services;
- manage containment errors of the physical layer through an independent bus guardian;
- be able to support the presence of a decentralised bus guardian on the physical layer (all topologies together), and be such that the protocol is independent of the use of a central bus guardian (optional bus guardian, no interference);
- provide the possibility of introducing new nodes into an existing system, without having to reconfigure the existing nodes; the configuration information must follow a 'need to know' philosophy via a specific, intrinsic device.

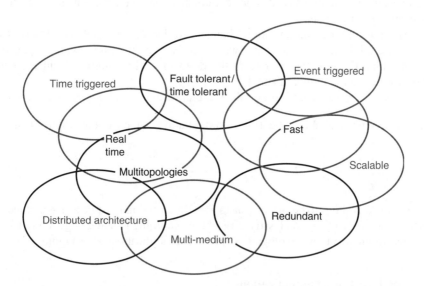

Figure 4.2 Global view of problems to solve

4.4.2.2   In Terms of Topology

To provide flexibility of vehicle design from high range (particular models) to mid-range (design of platforms), FlexRay must support communications which are implemented using topologies of the following types:

- single channel;
- dual channels;
- bus, bus with stubs;
- passive and active stars;
- multiple stars, with the possibility of sub-buses;
- and mixtures of all of them.

4.4.2.3   In Terms of Security Requirements

For systems of 'all controls by wire' ('X-by-Wire') type, FlexRay must:

- be able to manage redundant communication;
- provide deterministic access to the network (synchronous redundancy);
- avoid collisions for access to the bus;
- have the possibility of redundancy with variable geometry (single and double channel, mixed system);
- act according to a 'never give up' strategy, and thus ensure that all unavailable communication systems systematically inhibit distributed salvage mechanisms;
- be such that restarting a node in a system in operation is not limited to restarting its communication, but implies many more things;
- maintain communication for as long as communication between other nodes is not compromised.

Maximum security can only be reached by the combination of hardware (dual-channel architecture, independent bus guardian, fault-tolerant central processing unit (CPU)) and mechanisms included in the protocol and the application support.

It is also necessary to ensure:

- robustness of the system against transient faults and external radiation;
- a minimum of radiation externally to the systems, electromagnetic compatibility (EMC) protection, and so on.

4.4.2.4   In Terms of Application Requirements

FlexRay must ensure that the application:

- is always entirely responsible for all decisions to be taken in terms of security or availability of the network;
- always takes the final decision;

- always has control of the communication system – and not vice versa!
- maintains reception for as long as possible, since stopping communication is a critical decision which must also be made at application level;
- at communication level, can provide different operating modes; that is:
  - normal or continuous operation,
  - operation in dedicated degraded mode: warning (continuous operation but with notification to the host node), reduced operation with errors (transmission stopped and notification) and fatal error (operation stopped, all pins and the bus guardian go into 'fail safe status');
- is such that the nodes can be configured to survive for several cycles without receiving communication frames.

These general parameters and requests make it possible to address directly the future requirements of three classes of application which today are not covered by CAN or by other existing protocols. These are:

- **Class 1**: communication with high pass bands;
- **Class 2**: communication with high pass bands and of deterministic type;
- **Class 3**: communication with high pass bands and of deterministic, redundant type.

They make it possible to foresee, within 10 years, new network hierarchies being used in industry in embedded applications of FlexRay type, in the three application classes described above: use of the FlexRay protocol as the backbone of the overall system, CAN as a sub-bus of FlexRay and LIN as a sub-bus of CAN.

IMPORTANT NOTE

The administration of the FlexRay Consortium expressly invites all persons and companies wishing to use FlexRay under the FlexRay registered trademark to contact the administration concerning certification of products, systems sold, systems used, and so on, because in this case they must obtain an 'Essential FlexRay IP' licence by becoming members of the Consortium.

It is also important to note that the FlexRay Consortium emphasises that the FlexRay system was designed and developed only in the context of applications for the automotive market, and that its specifications were never developed or tested for non-automotive applications.

Finally, to make this very clear, the companies which are involved in the specifications of FlexRay accept no responsibility for the consequences of using FlexRay systems, in particular for non-automotive applications.

A LITTLE HISTORY

Rome wasn't built in a day, and nor were CAN and FlexRay. These brief histories are a reminder.

CAN is Already Over 30 Years Old!
For information, the following list summarises the principal stages which CAN went through during the first 25 years of its life.

- **1983**: Start of development of CAN at Robert Bosch GmbH.
- **1985**: v1.0 specifications of CAN. First contact between Bosch and the silicon founders.

- **1986**: Start of standardisation work at ISO.
- **1987**: Introduction of the first sample of a CAN integrated circuit.
- **1989**: Start of first industrial applications.
- **1991**: Specifications of the extended protocol, called 'CAN 2.0'.
  - Part 2.0A – identifier in 11 bits.
  - Part 2.0B – identifier in 29 bits.
  - First vehicle – Mercedes Class S, equipped with 5 units communicating at 500 kbit/s.
- **1992**: Creation of the 'CiA – CAN in Automation' – group of users.
- **1993**: Creation of the 'OSEK' group. Appearance of the first application layer, CAL, of CiA.
- **1994**: The first ISO standardisations, called 'high and low speed', are completed. PSA (Peugeot and Citroen) and Renault enter OSEK.
- **1995**: Task force in USA with SAE.
- **1996**: CAN is applied to the majority of engine controls of numerous European high-range vehicles which are part of OSEK.
- **1997**: All the large silicon founders offer families of CAN components. CiA represents 300 member companies.
- **1998**: New sets of ISO standards around CAN (diagnostics, compliance, and so on).
- **1999**: Development phase of CAN time-triggered networks, TTCAN.
- **2000**: Explosion of equipment connected by CAN in all automotive and industrial applications.
- **2001**: Industrial introduction of CAN real time, time-triggered networks, TTCAN. Even the Americans and Japanese use CAN!
- **2008**: World annual production of about 65–67 million vehicles, with an average of 10–15 CAN nodes per vehicle. You do the maths!

FlexRay is Already Over 15 Years Old!
For information, the following list summarises the principal stages which FlexRay went through during the first 15 years of its life.

- **1995**: Start of the 'by Wire' concept at BMW and Robert Bosch GmbH.
- **1998**: Comparative study by Byteflight/CAN/TTP/TTP/C. First contact with the silicon founders.
- **2000**: Creation of the FlexRay Consortium.
- **2002**: Introduction to the public in April, when FlexRay was demonstrated in Munich.
- **2002/03**: Introduction of the first samples of protocol management integrated circuits in the form of FPGAs and line drivers.
- **June 2004**: Specifications of the extended protocol called 'FlexRay 2.0':
  - protocol part.
  - physical layer part.
- **May 2005**: Specifications of the extended protocol called 'FlexRay 2.1':
  - protocol part.
  - physical layer part.
- **November 2005**: Second 'FlexRay Day'. Introduction of the first microcontroller with a FlexRay 2.1 communication controller on board. The complete set of certified components exists: driver, active star, combined microcontroller and communication controller.
- **January 2006**: Final development tools on the market: Vector/Decomsys, and so on.
- **End of 2006**: First BMW vehicle (X5 model), equipped with several units communicating at 10 Mbit/s.
- **2007**: Final definition of the AUTOSAR layer for FlexRay.
- **2008**: Production of a second vehicle from the manufacturer Audi. All the large silicon founders offer component families.
- **2009**: Third manufacturer: DaimlerChrysler.
- **2010**: Release 3.0 of FlexRay.
- **2015–2016**: First introductions forecast by the French manufacturers, PSA and Renault.
- **2020**: FlexRay equipment present in all automotive and industrial applications.

5

FlexRay and Real Time

Whether a communication network system architecture is simple (for example point to point) or complex (for example distributed and distributed intelligence), a protocol and its communication application layers function correctly only if the hardware (silicon, integrated circuits, and so on) and the physical layer (medium and topologies) on which they are implemented support them correctly! This seems so obvious that some people sometimes forget it!

As you will discover as you progress through the chapters of this book, because of the structure of the data transport mode adopted by the FlexRay protocol, the latter does not include the concept of acknowledging transmission in the communication cycle/frame (the quality of the transport of the FlexRay data being indicated only by providing a cyclic redundancy check (CRC)). It is therefore unnecessary to concern ourselves with the problems caused by the concept of managing the propagation time of a round trip signal on the network within the duration of a bit, to be certain that the protocol can function, as was the case with CAN. On the contrary, the fact that the principle of access to the network by the various nodes is built around a concept of TDMA type implies that each of them can or must jump into its time slot on the move, hoping to hit it just at the right time and transmit its data frame there! Hitting it isn't too difficult, it's the 'right time' which is complex because, in principle, there is no reason for the nodes to be synchronised with each other. In fact, for one thing the individual clocks are not quite the same, for another there is no reason why the nodes should all be at strictly identical distances from each other, and therefore the signal propagation times should be strictly identical, and so on.

After these very basic considerations, we will now go on to consider in detail the 'Time' (with a big T) parameter, and all the associated concepts derived from it. In fact, concerning the FlexRay protocol, various aspects which are more or less linked to time must be resolved.

5.1 Physical Time

Obviously, the 'time' which passes is a physical concept, and can be measured using the 'universal time' scale, for example the classic Greenwich Mean Time, which is used as an absolute reference to which the whole world can refer, but that is often not very practical. Strictly speaking, we can and must refer to this reference, as we will do occasionally.

FlexRay and its Applications: Real Time Multiplexed Network, First Edition. Dominique Paret.
© 2012 John Wiley & Sons, Ltd. Published 2012 by John Wiley & Sons, Ltd.

5.2   Local Time

Now consider a specific node of the network. This is classically equipped with a micro-controller which is driven/timed by an on-board clock.

5.2.1   Local Clock

In general, this 'local clock' is driven by a quartz crystal of good quality, and usually has low tolerances, low temperature and time drifts (component ageing), and if possible it is insensitive to variations of power supply (see the 'FlexRay and time' box later in this chapter). This clock therefore creates a local time which is specific to the node under consideration. Additionally, if desired, it is easy to extract from this local clock sub-multiple frequencies (classically, simply by dividing the local clock), but also to create higher frequencies (integer multiples or otherwise), for example using phase locking devices of 'phase locked loop (PLL)' or 'fractional PLL' type, so as to have faster local clocks which, for example, are intended for use with signal oversampling techniques.

Example

Table 5.1

Node i	Quartz (MHz)		Local clock (MHz)
	40	Divider: 2	20
	20	PLL × 4	80

This local clock lives its life independently of the rest of the world. Obviously, each node of index i of the network does the same on its side, and everyone knows that once a quartz crystal is soldered onto a printed circuit board, it is difficult to make it change the value of its frequency! In short, it's fixed! Also, however it happens, on a network with several nodes, for various reasons (temperature, and so on), after a few moments all the clocks diverge to some extent from each other and from their initial values – even after being synchronised initially!

5.2.2   'Clock Tick' and Microticks

5.2.2.1   'Clock Tick'

For reasons which we will explain later, it is necessary to create, locally in the node, a clock called 'Clock Tick' which is derived directly from the local clock. As indicated in the previous section, relative to the local clock its value can be lower, obtained by division, or higher, obtained using a PLL.

> **IMPORTANT COMMENT**
>
> The job of this local clock – 'Clock Tick' – and the specific duration of its period is to act as a physical basis for other local functional units of the node. But we will show below that its intrinsic value is not the relevant value with respect to what will be used to define the duration of the logical bit.

5.2.2.2 Microticks

'Microticks' are created locally, at the node itself, from the above-mentioned 'Clock Tick' clock, which is itself derived from the actual (quartz) local clock of the node.

By definition, the time value of a period of this new clock is called 'microtick' (μT) . This quantity forms the finest time hierarchy of the FlexRay protocol.

> **COMMENT**
>
> If the local clock of the node oscillates at a high enough frequency, the μTs can be generated directly by dividing the quartz clock.
>
> If this solution cannot be applied (power consumption of the microcontroller on the node too high, and so on), as was mentioned above, on the microcontroller which manages the FlexRay protocol there is often a PLL device to raise the frequency which is dedicated only to the FlexRay part (for example starting with 20 MHz, raising it to 80 MHz), and then to divide it to obtain the time values of the μTs which the application will require.

The durations of the μTs are thus, by definition, integer multiples or sub-multiples of the period of the clock of the local microcontroller of the CPU on the node. Each node i thus creates its own μTs, the time values of which – the durations, $μT\_i$ – cannot, because of their structure, be affected or altered by any external synchronisation mechanism. Using an image, it can be said that the μT is constructed using an immutable little electronic mechanism, which is specific to the node, but that obviously its value, being completely linked to the value of the local clock, will be directly influenced by, and only by, all the tolerances and drifts of the local oscillators.

5.2.3 In Practice

The μT – a time value which is linked directly to the value of the 'Clock Tick' – is the smallest FlexRay time unit. At the level of a node of the network, the granularity/ resolution/fineness/precision of the measurements of the time differences in FlexRay is therefore linked to this value.

Obviously, the smaller the duration of the μTs, the finer the granularity will be.

Example

Table 5.2

Microcontroller part of node				FlexRay part of node		
Quartz of node (MHz)		Local clock (MHz)	Local clock period (ns)	Tick divider	Tick clock (MHz)	μT period (ns)
40	Divider: 2	20	50	2	10	100
20	PLL × 4	80	12.5	2	40	25

For a conventional FlexRay network, the 'Clock Tick' is usually chosen to be 80 MHz (that is, a period of 12.5 ns), and two ticks are taken as forming a μT, giving it a period of 25 ns.

Almost to the last comma, it is self-evident that all the nodes on a single network therefore have μT values (μT_i) which are significantly different, representing the specific granularities of each.

The μT is the distinctive local time unit of a node. It is a local constant of the node.

COMMENT

Let us unveil one part of the mystery: the μTs will be used for creating macroticks (MTs), and later, with the latter, to assist the device for synchronisation between nodes.

After approaching a local aspect of time, let's go on to a global aspect!

5.3   Global View at Network Level – Global Time

Before going any further, we would like to ask you to be very attentive to some lines and paragraphs that follow, because they are fundamental to assimilating the concept of the FlexRay protocol well.

5.3.1   Concept of Global Time

So that the system functions correctly, sooner or later it will be necessary to put together in the same bowl of time all the participants which are present in the network and have functional reasons for working together (this group of nodes is called a cluster[1]). When this has been done – using synchronisation devices which will be explained in Part D – the cluster of this network will function under a common, global time unit called 'Global Time'.

'Global Time' will thus represent the general, common view of the 'time' parameter within this group of participants (cluster). As we will show, it should be noted that the FlexRay protocol as such has no Global Time reference or absolute time, but that each node has its own local view of Global Time (this sentence should be reread at least three times!).

Explained in this way, Global Time thus seems to be an abstract, rather immaterial entity ... which it is!

There is thus a latent conflict and a profound contradiction between, on the one hand, the concept of local time which is specific to a node, with its immutable values of μT,

---

[1] The generic term 'cluster' often causes confusion. To give you a more precise idea of its meaning, you should know that on the FlexRay network (as on others), numerous (30, 40 or more) nodes can be connected simultaneously: engine, gearbox, all the components which are somehow related to the suspension and road-holding of the vehicle, all the components which are related to the brakes, the air-conditioning, the comfort of passengers, safety, and so on. Some have nothing to do with others, but ... It is certain that the gearbox is interested in the engine speed, but who can prove that the engine is interested in the sunroof controls? On a single network or communication medium, some nodes can be grouped functionally (but geographically) into clusters, with functional properties to be shared and closer time constraints. Finally, a single network usually supports several clusters of nodes.

as described in Section 5.2.2.2 about microticks, and on the other hand the concept of a global, flexible and adaptable synthesis of individual times, from which the value of the Global Time of the network will be derived.

With the aim of creating a buffer between these two entities – local time and Global Time – an entity called the 'macrotick' has been created, with the function of being a network constant and the distinctive time unit of the network.

5.3.2 Macrotick (MT)

Globally, over the extent of a set of nodes (cluster) within the network, it is important to have only a single, unique concept of time, 'Network Wide Global Time' – simplified to Global Time – the value of which will be expressed in terms of 'MTs'.

By definition, the 'MT' is a quantity of time which concerns a particular set of network participants ('cluster-wide'), and can then be used to construct the duration of the bit ('bit time') for use by this set of participants. To explain this concept of Global Time in more detail, you should also know that FlexRay specifies that:

- successive communication cycles are numbered as $(2n + x)$ (see below);
- whatever happens, each communication cycle must have the same whole number k of MTs:
- duration of a FlexRay cycle $= k$ MT, where k is an integer. k is a constant of the global network.

Consequently, the MT is the smallest unit of global time granularity of the network. **The MT is the distinctive time unit of a network.**

If it is desirable that the time unit is constant, and given that k is also a constant, this means that:

- the duration of the MT must be adapted as a function of the other components/ parameters, to satisfy the equation above;
- the entity forming the Global Time of the network is formed by a pair of values: firstly the value of the cycle counter – Cycle_Count – and secondly the value of the MT counter – Macrotick_Count.

In parallel, for a specific node i of the network, we can establish the value of the ratio n_i which links the respective durations of MTs and μTs:

$$n\_i = \text{duration of an MT} \,/\, \text{duration of a } \mu\text{T}$$

By definition, the value n_i is a local value of a node, since it belongs to the controller i. In fact, the durations of μTs are (locally) specific to each node, and not to the whole network; also, in general, the values of these ratios are not integer values! As a corollary, this parameter n_i also represents the number of μTs per MT (by default, this value is part of the initial configuration parameters of a controller located in node i).

> **NOTE**
>
> Later, particularly in Part D about synchronisation, we will show that the value of this ratio is influenced by the clock synchronisation mechanism.

It is also important to distinguish:

- the abstract value of the MT as the time unit of the network (for example 1 μs);
- the local view of the duration of the MT during a communication cycle;
- the duration of a specified local MT.

5.3.2.1   At Network Startup

At the initial startup of the whole network, without knowing how the partners will react, the local value of the initial duration of each of the MTs of each node consists of a whole number of μTs, for example for node i:

$$MT = n\_i \times \mu T\_i$$

Why make it complicated when you can make it simple!

Given that we indicated above that the duration of the μTs, μT_i, is specific to each node as a function of the frequency of its own local oscillator/clock/(PLL) and the content of its internal prescalers, the corresponding durations of the MTs will also be slightly different from node to node. Thus, each node builds its own local duration of the communication cycle. An example of this scenario is shown in Table 5.3.

At this stage, we now have a local view of time at the level of a node, and a global view of a global time at network level. Figure 5.1, using Russian dolls, shows how these values are nested in each other.

Starting from the bottom of the figure – the local clock – it should be possible to construct the duration of the bit: starting from the top – Global Time – likewise. But, sticking strictly to the statements we have made so far, we won't obtain the same value!!! That seems careless! The value of the bit time is thus sandwiched between the two philosophies of local time and Global Time! So let's go on to examine the inside of the sandwich!

Table 5.3   Local duration of communication cycle

Node name	Local clock (MHz)	On-board PLL clock	Prescaler value	Duration of granular μT_i	n_i of μT per MT	Duration of MT (n_i × μT_i)	Number of MTs per cycle	Initial cycle time
A	20.010	With 80	–	–	–	–	5000	–
B	24.990	–	1	24.999	40	0.9999 μs	5000	4.9999
C	80	Without	1	12.5 ns	80	1 μs	5000	5 ms
D	16	With 32	–	–	–	–	5000	–
E	40	–	–	–	–	–	5000	–
F	–	–	–	–	–	–	5000	–

The global time

The cycle

The segment

The slot

The frame

The byte

The bit

The macrotick

The microtick

The clock

Figure 5.1 The values are nested in each other

5.3.3 And the Bit Time – What's Happening to it Inside There?

In principle, the bit time is a time value which is uniquely important for the abstract representation of the logical value, '1' and '0', of the bit.

All the network times from which the bit time is derived must therefore be calculated by the local node. This calculation must take account, on the one hand, of the fact that the local time of the node is mechanical, and on the other hand, of the fact that whatever happens, the cycle duration will include a constant number k of MTs.

Let us put our cards on the table. The MTs, with the help of the μTs, will organise themselves to ensure a first presynchronisation among the local clocks and the physical signal which is present on the network. In fact, as we shall show in Part D, as the network operates the value of the MT is calculated or adjusted cycle by cycle, using an algorithmic synchronisation procedure. The construction of the MT is thus not linked to a simple story of electronic mechanics as in the case of μTs, but follows a clever calculation.

Additionally, the number of μTs per MT can be different from one MT to another within the same node. Although at startup any one of the MTs consists initially of a whole number of μTs, the mean duration of all the MTs of a whole communication cycle can be a non-integer value; that is, it can consist of a whole number of μTs plus a fraction of a μT. It is these time adjustments, made by calculation, of the value of MTs – which are themselves directly linked to the μTs, which are in turn linked to the frequency of the microcontrollers of the CPUs – which provide the synchronisation between the signals which are present on the network and the microcontrollers.

Let us look at that in detail.

Each (local) controller must manage three relevant values/parameters, k, n_i and d_i:

- k: the number (integer) of MTs per communication cycle; the value k is a constant of the network.

- n_i: this parameter represents the number of μTs assigned by default to an MT at the initialisation phase (it is part of the configuration parameters of a controller located at node i).
- d_i: in normal operation, after the network has succeeded in determining a Global Time as a function of the time performances of each of the participants, this value represents the number of additional μTs (plus or minus) that each node i will adjust per communication cycle.

Consequently:

- in principle, whatever happens, a communication cycle consists of k (integer) MTs;
- because n_i is also an integer, in an ideal world, if all the n_i of all the controllers are the same, a communication cycle should equal $(k \times n\_i)$;
- unfortunately our world, though beautiful, is not ideal, and on the one hand all the n_i of all the controllers are not the same, and since k is a constant of the global network, on the other hand each node must adjust (up or down) a local (total) number d_i of μTs, in such a way that the output communication cycle always remains equal to k MTs whatever happens! Consequently, in normal operation, for each of the nodes i the equation below is always, at every moment, strictly true:

$$\text{one communication cycle} = k \text{ MTs} = [(k \times n\_i + d\_i] \text{ μTs}$$

It is obviously possible to deduce from this equation that the mean duration of an MT is $[n\_i + (d\_i/k)]$ μTs, and that by intelligent manipulation of the number d_i of additional μTs, it is then possible for each node to raise or lower the value of its own cycle time, to adjust itself to a value of Global Time which is common to all the participants in the network. Finally, and as a reminder:

- the value of d_i (algebraically) additional μTs is a local value of the controller of node i, and we will show in Part D how this value is influenced via the synchronisation mechanisms of the nodes;
- it is also the responsibility of the hardware on the nodes under consideration to distribute the d_i (algebraically) additional μTs uniformly to the k (integer) MTs which form the communication cycle.

5.4   Summarising: Time and its Hierarchies in FlexRay

For each node/unit/module which is part of the network and includes a microcontroller, there are potentially several time units to be managed. Figure 5.2 is a reminder of the four different abstraction levels which we have mentioned; that is, the cycle, the slots and minislots, the MT and the μT. As a final reminder, an MT consists of a whole number of μTs, and a cycle consists of a whole number of MTs. The duration of the μT is linked

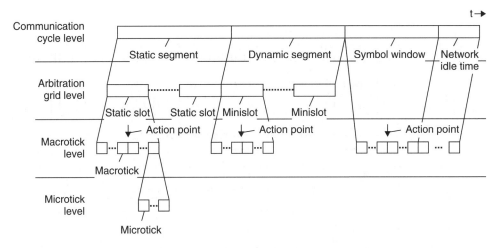

Figure 5.2   Time hierarchy defined by the FlexRay protocol

directly to the oscillator of the communication controller, and can therefore vary according to the different controllers.

COMMENTS ON FIGURE 5.2

- From now on, make the effort to read this figure from top to bottom and from bottom to top – depending on whether you want to be a *network* or a *local node*!
- The top of the figure is linked to local time. The bottom is linked to Global Time. After the clock synchronisation sequence (see details in Part D), there will be, in fact there is, an effect of the top of the bottom on the bottom of the top (You are following this, aren't you?).

From top to bottom: as seen from the local node:

- the local clock of the controller of a specific node/participant;
- a parameter derived directly from the latter is called 'microtick'.

From bottom to top: as seen from the network as a whole:

- the duration of an observable cycle on the network (can be measured using an oscilloscope!);
- the basic element of the cycle – the 'MT' – which makes it possible to construct its duration.

As a summary and conclusion, Figure 5.3 shows the set of different time references with which we shall be concerned in a network of FlexRay type. Table 5.4 shows an example of a hierarchy of time values.

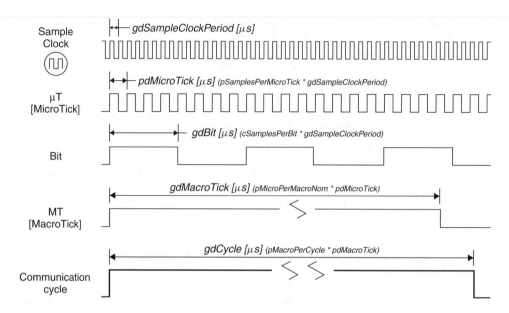

Figure 5.3   The set of different time references

Table 5.4   Example of hierarchy of time values

Parameters	Values	Units	Range of possible values
Quartz oscillator of node	20	MHz	–
Multiplication factor of PLL	4	–	–
Clock Tick oscillator	80	MHz	20/40/80
Period of oscillator	12.5	ns	50/25/12.5
Number of clocks per microtick	2	1/µT	1/2/4
Duration of microtick	25	ns	12.5/25/50/100
Bus speed	10	Mbit/s	2.5/5/10
Bit duration	100	ns	400/200/100
Number of microticks per bit	4	µT/bit	–
Samples per bit	8	–	8
Communication cycle	2	ms	0.010–16
Number of macroticks per cycle	2 000	MT	10–16 000
Duration of macrotick	1	µs	1–6
Number of bits per macrotick	10	–	–
Number of microticks per macrotick	40	µT/MT	40–240
Number of microticks per cycle	80 000	µT	640–640 000

Example

FLEXRAY AND TIME

This section is principally aimed at presenting and emphasising the various electrical and technological parameters which have direct and indirect effects and consequences on the time values of the signal:

- **As a pure value, at a given temperature and at 0 hour** – Although the FlexRay bit rate is perfectly and ideally defined as 10 Mbit/s (that is, a bit duration of exactly 100 ns) it is not at all certain that the local clocks of all the nodes of the network are perfectly capable of initially constructing a bit of which the duration is exactly 100 ns. For example, this can be because of tolerances – for a quartz of good quality, they are generally around a minimum and maximum of the order of ±250 ppm (9.999750–10.000250 Mbit/s) to obtain the best results in terms of topological flexibility and minimal residual asymmetry of signals – and drifts over time of the clocks which are driven by their respective quartz crystals.
- **Jitter[2] of the local clock** – The duration of the incident bit is often measured and validated using an oversampling technique, the frequency of which (higher than the node's local clock, for example 80 MHz) is obtained using a PLL on the basis of the output frequency of the quartz (for example 20 MHz). In this case, it is necessary to take account of possible jitter caused by very slight instabilities (noise, interference, and so on) of the phase control loop. To fix an order of magnitude, conventional applications usually allow for a jitter value of the order of ±0.5 ns. Also, the jitter from each edge of the signal sampling clock must not exceed ±2.5 ns, and in this case the cascade of clock jitters must not exceed ±160 ps.
- **Relative and absolute phase** – Additionally, even supposing that all the clock values are strictly identical, the fact remains that they must all be locked into phase relative to each other, so that they are no longer merely synchronous but become totally isochronous. As we shall see in Part D, which is about synchronisation between nodes, these various problems lead us to consider very precise concepts of corrections of phase (offset) and rate.
- **Drift** – To the problems of precision, tolerance and jitter mentioned above, the classic problem of time drift of components influencing the intrinsic qualities of the clocks must be added. This drift is principally caused by effects of temperature and ageing.
- **Temperature** – All the nodes are subject to classic variations of ambient temperature (−40, +70 °C), and also, for simple functional reasons, do not have thermally identical geographical positions in a network (under the bonnet of a motor vehicle, external on the wing of an aircraft, and so on). In general, this implies that it is necessary to allow for a more extended temperature range, for example from −60 to +125 °C.
- **Time (ageing)** – Similarly, at the same temperatures, there are also drifts in time (over several years) because of ageing phenomena or other constraints of certain parameters, some of which have a direct, critical effect on the 'time' parameter.

All these variations, precisions, tolerances, jitters, phases, drifts imply that, sooner or later, efficient devices for synchronising or resynchronising clocks must be defined and constructed (in a similar spirit to what was designed for the CAN protocol, and identical in broad outline), to make it possible to extract a received bit precisely and reliably.

---

[2] Jitter: rapid and/or erratic variation of the frequency around the central value of the oscillator (not necessarily the nominal value).

6

The FlexRay Protocol

This chapter presents the broad outline of the operation of FlexRay. It is subdivided into several parts, respectively concerning: transmission channels, cycles and segments, slots and communication frames, constructing and coding frames, the transmission start sequence (TSS), frame start sequence (FSS) and byte start sequence (BSS) systems, and the concepts of action point (AP), byte, bit and local clock.

6.1 History

On 30 June 2004, three reference documents were made public on the website of the FlexRay Consortium; these concerned version 2.0 of the protocol, the physical layer and the preliminary bus guardian. The final versions 2.1 (March 2005) and then 2.1 A and B a little later give some additional details and minor modifications, and version 3.0 '2010 version' puts the finishing touches by providing some additional points and details.

If you want to, you can download these documents from the Consortium's website (FlexRay.com). The original document which describes the FlexRay 2.1 protocol is over 300 pages. But be careful, it's quite forbidding and difficult to digest without bicarbonate of soda, particularly if one is not privy to all the little secrets of the gods who were called on when the protocol was developed ... and there would have to be at least 1000 more pages to explain in detail how it works.

Meanwhile, to make your mouth water, the sections which follow were designed to give you a detailed survey of this concept, and not to do this exhaustive study. For fans of detail to the nearest fraction of a bit, all you need is one website address and a good dose of courage.

While you wait, here, quickly, are some important points from these documents.

6.2 General – Channels, Cycles, Segments and Slots

Before going on to the content of the FlexRay protocol, let us begin by indicating the general philosophy of its operation, which is fundamentally different from that of CAN and other protocols which are used today by industry in this application field. This will take several paragraphs, but it is necessary in order to understand all the subtleties.

FlexRay and its Applications: Real Time Multiplexed Network, First Edition. Dominique Paret.
© 2012 John Wiley & Sons, Ltd. Published 2012 by John Wiley & Sons, Ltd.

6.2.1   Philosophy of the Protocol

First, by its structure, FlexRay was designed to provide a communication system which can function on one communication channel or simultaneously on two, and in which there can be no collision for access to the medium; this means that none of the nodes will carry out arbitration on the transmission channel, and that collisions must not occur in normal operation.[1] The physical layer provides no means of resolving these collisions, and if necessary it is therefore the application layer which must take responsibility for managing these problems.

To give very great application flexibility to the system, the system must be able to:

1. Function in real time; that is, communicate at precise, known instants, during a known maximum time, being certain of being the only one at this instant on the physical communication medium, and thus also with no possibility of collision.
2. Communicate event by event with a useful rate of variable information as required, thus to take a certain communication time which is unknown, but with a known maximum time limit.

The two points described above are somehow conflicting and deeply contradictory. There is therefore a circle to be squared – and evidently this is just what has been done, more or less to π!

To do this, in the same spirit as TTCAN, FlexRay proposes communicating by carrying out communication rounds called 'communication cycles', in which access to communication is implemented 'synchronously' and 'asynchronously' in slots of time dimensions which are well defined by the system designer (you, me, etc.).

In short, the time for which these cycles last is subdivided into equal time slots and minislots (see further on in the text) which belong exclusively to dedicated CPUs, thus enabling them to transmit their data. Exclusive (to the CPUs) assignments of these communication slots are carried out 'offline'; that is, during the system design phase, which in principle eliminates all competition for access to the network and other side effects during the normal operational phase of the network, called 'online'.

Of course, it is the responsibility of the designer or whoever is responsible for designing and managing the network to choose these values well. This principle of access to the medium by observing predefined time slots (TDMA, see below) eliminates structurally any possibility of message collisions (collision avoidance).

That being defined, FlexRay includes two paradigms[2]:

1. the first ordered/initiated by time (time-triggered);
2. and the other driven by external events (event-triggered).

To do this, within a communication cycle, it is necessary to create two completely distinct areas, one called the 'static segment' and the other called the 'dynamic segment',

---

[1] However, collisions may occur during the startup phase of the protocol on the transmission channel.

[2] Paradigms (definition from Larousse dictionary, translated): model of declension or conjugation. Example: the French verb 'finir', taken as an example or model of conjugation of verbs of the second group. It should be noted that the term 'paradigm' is often used in and about FlexRay about the numerous declensions which exist around the fundamental principles.

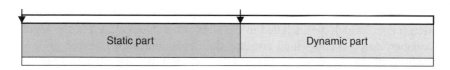

Figure 6.1   Static segment and dynamic segment of communication cycle

which are strictly dedicated to (1) and (2) above respectively – and that will almost do the trick (see Figure 6.1).

6.2.2   Hierarchy and Overall Form of FlexRay Communication

The FlexRay communication protocol is structurally constructed according to a parallel dual interleaved hierarchy of Russian doll type (see Figure 6.2):

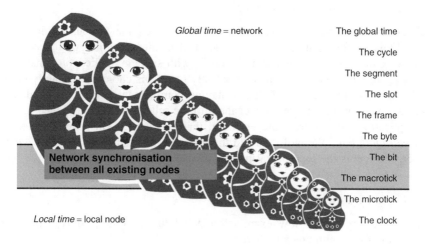

Figure 6.2   Parallel dual interleaved hierarchy of the FlexRay protocol

Table 6.1   Global view of FlexRay timing

Time	Communication				
	Channel				
Cycle time	Cycles				
Macroticks	Segments	Static	Dynamic	SW	NIT
	Slots	Slots	Minislots	Symbol	
	Frames	Static	Dynamic		
	Bytes				
Bits	Bits				
Microticks					
Clock ticks					
Local clock					

- The first, of functional type, concerns the encapsulation of bits and bytes in communication frames, which are included in the slots which are included in the segments of the communication cycles which circulate on a communication channel.
- The second, of time type, is the one in which we will mention the terms Global Time, macroticks, microticks, local clock, and so on.

To have a global view of what awaits you, the majority of the elements of the FlexRay protocol are shown in Figure 6.3, pompously called 'all in one'. The whole goes from the top (network level) to the bottom (level of the local microcontroller), from the largest (communication cycle) to the finest (microcontroller clock).

6.3 Channels and Cycles

Let's go over this magnificent figure in detail!

6.3.1 Communication Channel(s)

The communication channel represents the link between the various participants of a network. FlexRay forces us to design (at least at semiconductor level) systems which support two communication channels, A and B, with specific application constraints and flexibilities, but which can also function on only one. This choice was mainly guided by the fact that it provided the possibility of having complementarities, or redundancies, or withdrawals of information from one channel to the other, and that it opened the door to future applications which are more electrical than mechanical, in general called 'X-by-Wire' (see Figure 6.4). We will return to these points in detail in Chapter 9, which more particularly concerns the topological problems of networks.

6.3.2 Communication Cycle

Starting at the top, let us now examine in detail the structure, constituent parts and decomposition of a communication cycle.

Communication which is carried out using the FlexRay protocol is organised by 'communication cycles', which are recurrent and of equal duration.[3] In principle, the cycle duration is constant and linked to a 'Global Time'. We will describe later how that operates.

The recurrent communication cycles of the FlexRay protocol are numbered '$(2n+x)$' (see Figure 6.5). This deliberate '$2n+x$' numbering indicates that, sooner or later, it will be necessary to take account of the parity of the cycle number in the operation of FlexRay.

Let us examine briefly the general structure of the FlexRay communication cycle.

6.3.2.1 Its Constituent Parts

As shown in Figure 6.6, the structure of FlexRay communication cycles is subdivided into four distinct parts called 'segments', which are repeated cyclically at constant time intervals.

---

[3] All nodes have their own individual time bases, which are then synchronised with each other.

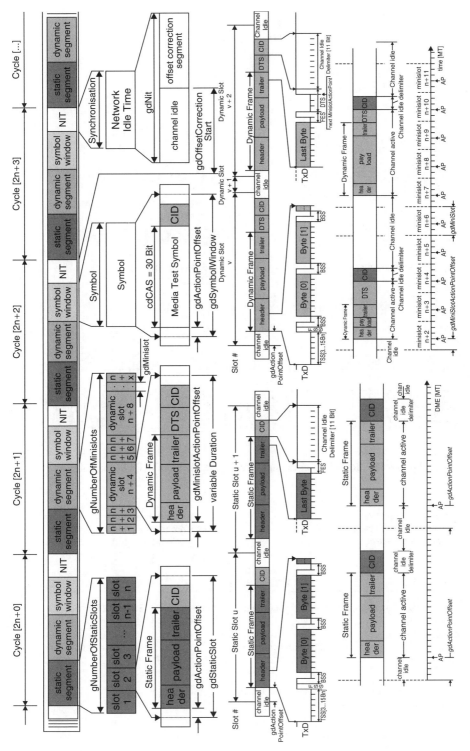

Figure 6.3 The FlexRay protocol

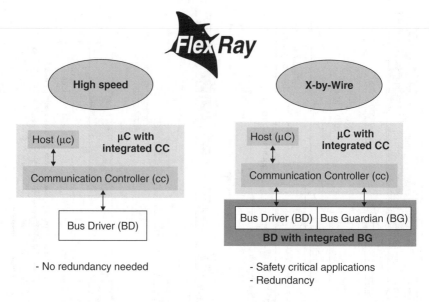

Figure 6.4 'X-by-Wire' applications

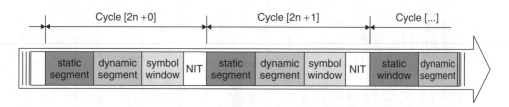

Figure 6.5 Recurrent communication cycles of the FlexRay protocol

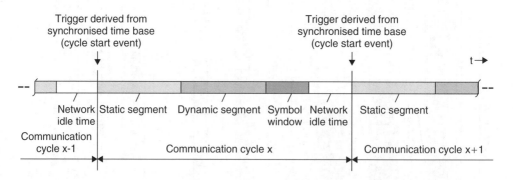

Figure 6.6 Structure of FlexRay communication cycles

Each cycle is made up of the following segments:

- ST, a 'static segment';
- SD, a 'dynamic segment', which may be optional;
- SW, a 'symbol window', which may also be optional;
- NIT (network idle time) to terminate the cycle; this is a time phase during which the network is in idle mode, and is therefore called 'network idle time'.

The communication cycle can have four quite distinct aspects according to the options which are chosen for the presence and subdivision of segments (see following paragraphs).

Additionally, the position of the boundary between the static and dynamic segments is left completely to the user's judgement, so that the user can take the most advantage for the application of the user's system. These facilities are part of the famous application flexibility which resulted in the name FlexRay.

6.3.2.2   Its Duration

Because of the desired aim – applications functioning in real time, and therefore fast – the time value of a FlexRay communication cycle must be fixed between 10 µs minimum and 16 ms maximum. Thanks to certain application tricks, it is possible either to apparently shorten the minimum cycle time (by certain slot repetitions) or to apparently increase the maximum cycle time by carrying out time multiplexing of cycles (see example a little later in this chapter).

> **NOTE**
>
> During the initialisation and/or wakeup phase, the communication cycle and its sequencing are initialised by the principal management node(s) of the network (see Chapter 15).

6.4   Segments

As we indicated above, the structure of FlexRay communication cycles is subdivided into 'segments', which are repeated cyclically at constant time intervals. There are four (see Figure 6.6 again):

- ST, a 'static segment', which is dedicated to deterministic, real time applications, with a known, determined bandwidth;
- SD, a 'dynamic segment' (optional), dedicated to event-triggered applications, subject to probabilistic management, with variable bandwidth;
- SW, a 'symbol window' (optional), specific to applications which use a bus guardian (for example of the type with a dual transmission channel and redundancy of X-by-Wire type);
- finally NIT, a phase called 'network idle time', during which the network is in idle mode.

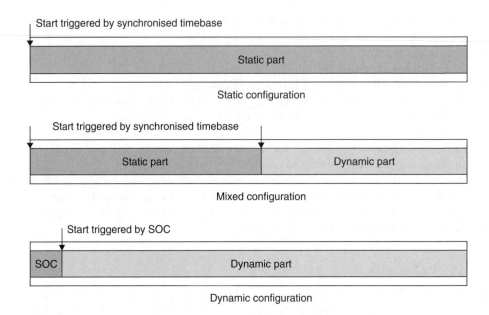

Figure 6.7   Possible applications

Because two of them are optional, Figure 6.7 shows all the possible variants of their applications, thus opening up very broad application fields. Again, this specificity and flexibility for an application is one of the numerous ingredients of this type which have supplied the 'Flex ...' part of the name of the protocol.

As of today, most applications are not (yet) of X-by-Wire type (and therefore without a bus guardian and the SW segment), and the operating mode which is most used is that which includes ST, SD and NIT. Users decide on the ratios of the durations of ST and SD within the communication cycle as they think fit.

COMMENT

Apart from particular cases which require special attention, and to avoid making this book too heavy, most of our explanations will be based on this last structure, ST, SD and NIT.

6.4.1   A Little Philosophy about Static and Dynamic Segments and Their Purposes

Before going any further, let us begin with some important reflections on the why and how of static and dynamic segments and their contents, slots and minislots.

As we indicated above, the aim of creating static and dynamic segments in the communication cycle is to square the circle; that is, simultaneously to provide and implement, on the one hand a communication system, for access to the medium, of deterministic, real time type with a known bandwidth, and, on the other hand, a system for access to the medium which is event-triggered (for example spontaneous events) with a variable/adjustable bandwidth. This is the reason that now leads us to consider in detail the

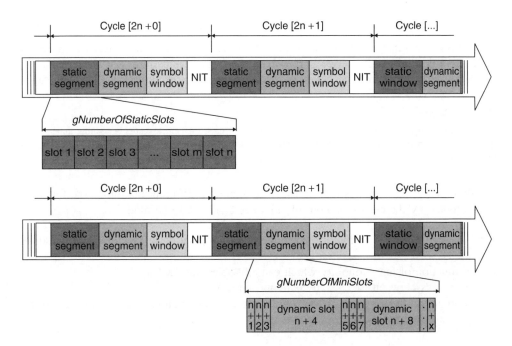

Figure 6.8   Slots and minislots

constituent parts and internal subdivisions of the segments, and the creation of 'slots' and 'minislots' and their respective functions.

6.4.2   Slots and Minislots

The static and dynamic segments described above are, in turn, respectively divided into time 'slots' and 'minislots' (see Figure 6.8). The purpose of these time 'slots' and 'minislots' is to transport the communication frames – static and dynamic respectively. We will, of course, return to their structure and content in detail in the course of this chapter and the following chapters.

When one wants to be able to use a system of real time type, that implies that previously the system designer has defined carefully what he or she wants to transport in terms of messages, and that it is then possible to define suitable time periods which are well fixed and well determined in duration and time position. These are the 'slot times', or 'slots' for short, of the static segment part of the communication cycle. Additionally, knowing that spontaneous events can also occur at any moment, it must be possible in principle to provide possible instants for starting slots for them in the communication cycle. This is the origin of 'minislots', which take their place naturally in the dynamic segment to solve this problem.

6.4.2.1   The Slots Philosophy and its Industrial Consequences

Before descending to the content of the slots, let us pause for a few moments to consider the philosophy which is hidden behind a 'slotted' structure.

6.4.2.2   Functional Configurations of Nodes and Clusters

Nowadays, the functions to be implemented to ensure that systems function well are distributed throughout the network. However, the respective activities of all the nodes on the network do not make it necessary to be interested in all the signals of all the messages.

SUMMARY

In a group of participants (cluster), it is unnecessary for each of them to know everything about all the messages which circulate on the network, and this minimises the overall memory size which is required for the configuration.

Additionally, throughout the life of a system, and on different platforms, the functions implemented by the electronic control units (ECUs) have to be modified

- by making a mixture or mixtures of functions from a set of several ECUs, for a subset of ECUs, or
- as a function of the variability which can exist between ECUs:
  - mappings between ECUs and changing messages;
  - reprogramming the network as a whole, while avoiding unnecessary coupling between message and ECUs.

6.4.2.3   Flexibility of Integration and Industrial Consequences

The principle of being able to have slots which are totally independent of each other for the duration of the static segment provides great flexibility and numerous application possibilities for manufacturers. In fact, since each slot of the static segment is assigned to a specific function of a node, and since it is no longer necessary either to regulate the possible collisions between identifiers (IDs) or to manage arbitration conflicts for access to the medium on the network between the frames sent by the nodes, it becomes thinkable for the system integrator to 'divide and rule'. In fact the system integrator can easily assign slot A to a given function, and assign implementation of it to supplier X, and similarly for slot C given to supplier Z. If the system manager/architect of the network has a sound view of the project, no-one else knows the interactions which exist between tasks A and C, and he or she can work serenely without suppliers X and Z needing to know or be in contact with each other, even if a crisis occurs, as could sometimes be the case when CAN was used. Additionally, since there is no longer any problem of conflicts between nodes for access to the network, when functions are integrated into systems, all the 'boxes' A, B, C, and so on can only fit one after the other into their respective slots of the static segment, thus providing an integration time which is shorter, less expensive and above all, in principle, without problems.

Figure 6.9 illustrates these possibilities (supplier/equipment manufacturer A, B, system integration and OEM (Original Equipment Manufacturer)).

Additionally, all well-designed functions A or C can easily be reused (the great industrial theory of 'reuse') in a new system, to create new vehicle platforms, and to make it easy to respond quickly to the expectations of the market – the other great theory of 'time to market' – because of even more reduced integration time.

All these advantages form the visible face of the Moon, where all equipment manufacturers see the good side.

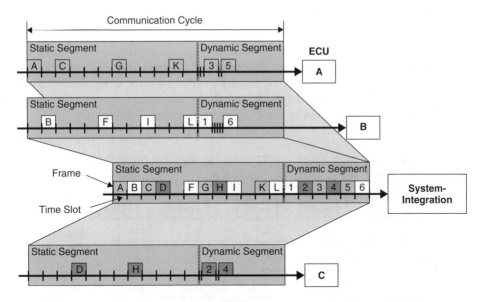

Figure 6.9   Integration of functions into systems

The hidden face of the Moon is more subtle. In fact, it is very easy for the system integrator to give the same specification of functions A or C to other possible suppliers, which, with the same performance and at lower costs, will take immeasurable delight in knocking your solution out of the way, knowing that their modules will go into the static slot under consideration, with no surprise or risk, as in 'plug and play'!

Now that you have been warned about the good and the less good, the advantages and consequences you get are up to you.

6.4.2.4   Technical Functions

The numerous different technical functions and the tasks related to them each require different communication bandwidths. The FlexRay format offers several possibilities for operation, for example:

- The possibility that a single node can transmit several messages (frames) in several slots of a single segment/cycle.
- The possibility of being able to start and close loops of distributed commands within a single communication cycle (for example to carry out a particular task within a single communication cycle). For instance, the question can be raised during slot B by the corresponding node, and the response can be given a little later by another node, for example during slot G of the same communication cycle.

Figure 6.10 shows the principle of response in the current frame:

- step 1: nodes 1, 2 and 3 send information in slots S1, S2, S3;
- step 2: nodes 2 and 3 are interested in the messages;

- step 3: the application calculates;
- step 4: the results of the application are given in slots A1, A2, A3;
- step 5: nodes 2 and 3 are interested in the results.
- The possibility of closing loops of distributed commands from one communication cycle to another. For instance, to carry out a particular task, the question can be raised during slot B by the corresponding node, and the response can be given a little later by another node, for example during slot G of another communication cycle.
- and so on.

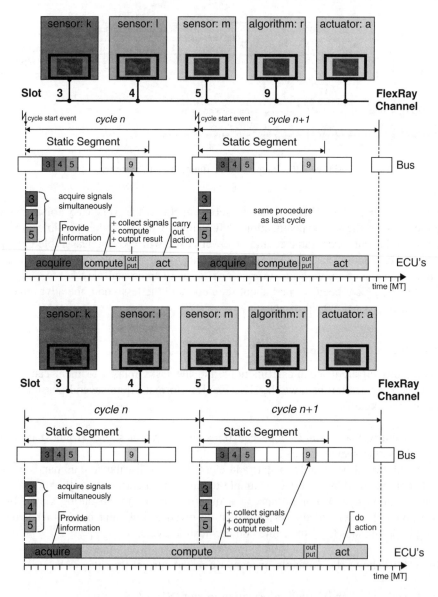

Figure 6.10 Principle of response in the current frame

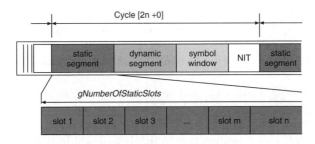

Figure 6.11 Static segment

Additionally, application protocols of the highest level (application layers of the OSI model) can be supported by the communication system:

- the consistency of data in an application is implemented by an application end-to-end agreement;
- multiple sending slots for a node make it possible to finalise an agreement within a cycle;
- the cadence of FlexRay nodes is determined on a principle of 'need to send'.

Now that's all settled, let's get back to our two subjects, firstly the static segment and its slots, and secondly the dynamic segment and its minislots.

6.4.3 Static Segments and Slots

As shown in Figure 6.11, by definition, the FlexRay protocol uses the name 'static segment' for the whole portion of the communication cycle during which access to the medium is controlled using an operating principle of static TDMA type, or global time division multiple access (GTDMA).

6.4.3.1 Purpose of the Static Segment

The purpose of this segment is to permit and ensure high-performance deterministic communication, to define precisely the semantics (meaning) of the state messages which are carried and to manage the distributed systems and the controls/commands of closed loops. As we will show below, the intrinsic form of this segment causes great advantages and benefits regarding the design and simulation of distributed functions and applications of real time, critical and safety type.

6.4.3.2 Structure and Time Subdivision of the Static Segment into Slots

For use, the static segment is subdivided into 'slots' (time periods, see Figure 6.12). Some rules define this subdivision precisely:

- all slots have the same duration;
- all communication cycles must include the same number of slots;

- each slot is identified by a unique number (slot number, unique ID);
- in the case of use of FlexRay communication in dual-channel mode, the formats of the slots on the two channels (duration, number, and so on) are identical;
- the starting instant of a slot is determined by the Global Time of the network (see below);
- via the above Global Time, the duration of the slot is also defined on the bases of groups of participants (clusters);
- only one node per slot is permitted to output;
- the IDs of the slots have a unique task, which is assigned in relation to the outputting nodes Tx;
- finally, it is necessary to use at least two slots of the static segment during the synchronisation procedure.

6.4.3.3   Implications and Consequences of These Rules

The slots include gaps (actually silences, see Figure 6.13) which are well known, well structured and begin and end at instants which are precise and mainly defined by application groups (clusters). All that is necessary is to assign these time slots by name to certain nodes or tasks, and then, by design, encroachment of communication between nodes is impossible. It is thus made possible to:

- implement a system in which access to the medium is distributed in time – TDMA or 'time division multiple access', also sometimes called 'time distributed multiple access';
- to avoid creation of collisions (contention), by structure and systematically – collision avoidance;
- to assign slots by channel for a transmission node (as we will see later, a single transmission node can communicate on the two channels in different slots);
- to reserve/assign specific slots to each node;
- to make the system real time, since the latency times are known because of the cyclical aspect of communication;
- for a bit of given duration, to know exactly the maximum bandwidth (a determined, known value by the initial design of the system; all the static slots have the same possible maximum bandwidth);
- also, in the static segment a maximum of 16 slots can be assigned per node.

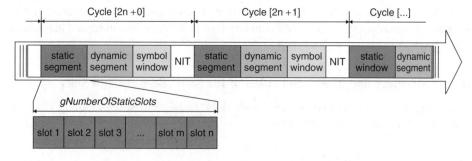

Figure 6.12   Static segment

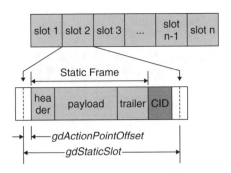

Figure 6.13   Gaps between slots

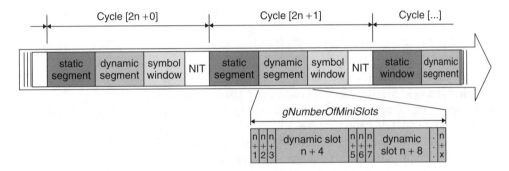

Figure 6.14   FlexRay dynamic segment

6.4.4   Dynamic Segments and Minislots

As shown in Figure 6.14, by definition, the FlexRay protocol uses the name 'dynamic segment' for the whole portion (optional, but very often present) of the communication cycle during which access to the medium is controlled using an operating principle of flexible time division multiple access (FTDMA) type, which we will now explain.

6.4.4.1   Purpose, Structure and Time Subdivision of the Dynamic Segment into Minislots

For use, the dynamic segment is subdivided in advance, offline, by the system designer, into 'minislots' (short time slots or periods) of identical (often short) duration, and thus starting and ending at precise instants (see the detail of this minislotting in Figure 6.15).

In fact, in contrast to the static segment described above, which is divided into (large) slots of equal, defined duration (because it was desirable to know precisely the instant at which one wanted a node to be able to access the network), the total duration of the dynamic segment is subdivided into numerous slices, short equal time periods called 'minislots', in expectation of ... mystery! A little suspense!

The time positions of these minislots within the dynamic segment are numbered in the segment itself, and also in relation to the full cycle, and their assignments are defined

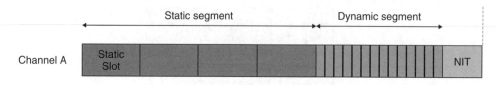

Figure 6.15 Minislotting

offline by the system designer. They represent 'possible instants' at which, depending on the intended applications, a node can start a communication element (subject to certain conditions which we will explain in Chapter 14). When, following an event which is untimed, spontaneous, to be transmitted (so outside a rigid timeframe as represented by the static segment described in the previous section), a minislot is used, it changes its name and thus becomes a 'dynamic slot'.

Some rules define and govern minislots:

- by definition, all cycles include the same number of minislots;
- their durations must be identical on the two transmission channels;
- the sequence of access to the network during the dynamic segment may be different on the two communication channels;
- the unique task/function of the output messages (Tx) is strictly linked to the node, chosen dynamic slot and given channel parameters;
- the start of the dynamic segment is linked to the different modes of using segments; that is:
  - in mixed mode – that is communication cycles during which both static and dynamic segments are present – the start takes place according to Global Time,
  - in the mode in which only the dynamic segment is present, the start is initiated by SOC (start of cycle)
- when a communication element has started, at the time of a determined minislot number so-and-so, that becomes a dynamic slot.

6.4.4.2 Implications and Consequences of These Rules

Obviously, the duration of a dynamic slot is essentially variable according to whether the transmitting node is more or less chatty (obviously within limits!). Consequently, in principle, any node can jump – at its event-triggered rhythm – onto the moving train represented by the dynamic segment. Obviously, there is no reason why two nodes wouldn't want or need to start a communication element at the same time! At least one hopes there isn't.

The reality is more complex. To avoid conflicts for access to the medium during the dynamic segment, it was decided also to assign a unique number 'ID Dynamic' to the various network nodes which are capable of transmitting (via the value of the 'frame ID' contained in the header of the message which the node wishes to transmit during the dynamic slot). Consequently, if two or more candidates are eager to start during the dynamic segment, an 'arbitration' phase (actually a decorum phase, see Chapter 7 for the part about conditions for access to the network medium) will be carried out according to the hierarchical level (and other much more complex trifles which are explained later) of the unique ID which is assigned to each node, and a single node will access the network without there really being a conflict.

6.4.5 Summary

The dynamic segment makes it possible:

- concerning data:
  - to ensure a space of time during which a transfer of data of event-triggered type and spontaneous messages can be carried out;
  - to have a communication mode which is limited in duration and bandwidth;
  - once the minislot has been started by the node which has succeeded in accessing the medium, it can if it wishes (under the control of the system designer) vary the duration of its dynamic slot according to the quantity of data which it wishes to transmit, and according to the requirements which it will have to manage;
  - to have a system of variable bandwidth, since the message can have a different duration according to requirements;
  - to carry out transmission by time bursts of information;
  - to facilitate management of (event-triggered) diagnostic information;
  - in general, to transfer ad hoc all kinds of messages.
- concerning access to the medium:
  - to have easy access to the medium, distributed over time, of FTDMA type;
  - throughout the duration of the dynamic segment, to access the medium on the basis of priorities assigned to the nodes which have data to transmit (the lower the binary value of the ID, the higher the access priority);
  - to make it possible to implement a hierarchy of access to the network medium according to the value of the ID of the output frame, so that data collisions and competition between transmissions are no longer possible.

COMMENT

The precise instant at which a dynamic slot starts is linked to the unique ID of the minislot which was assigned to it and the durations of the messages which are present in the preceding minislots of the dynamic segment. In fact, if the available bandwidth in the whole dynamic segment is less than the sum of all the frames which must be transmitted during it, those which have the number ID with the highest binary weight wait for the next transmission possibilities in the next communication cycles.

Summarising, using these two types of segments, which are defined offline, FlexRay supports a static segment for messages of time-triggered type, and a modifiable (scalable) number of spontaneous messages of event-triggered type in its dynamic segment.

6.5 Communication Frames

Let us now go on to the *logical content* which will be contained in the slots and minislots.

6.5.1 Overview of Frames

FlexRay communication frames are transmitted in the same way in static and/or dynamic segments, in static and dynamic slots respectively, as shown overall in Figure 6.16.

The slots of static segments and minislots of dynamic segments are occupied respectively by static and dynamic communication frames. These have a family resemblance, like twins ... with some exceptions, as usual!

6.5.2 Common Constituent Parts of Static and Dynamic Frames

The constituent parts of the static and dynamic frames mentioned above have many similarities, which we will now examine (see Figure 6.17).

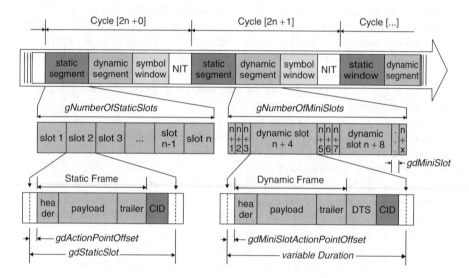

Figure 6.16 FlexRay communication frames: transmission

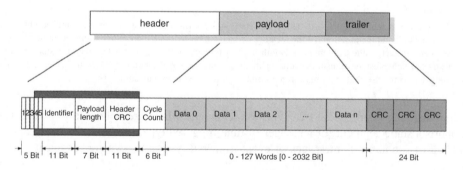

Figure 6.17 FlexRay communication frames: constituent parts

They are both principally composed of a trio of large, very distinct fields:

- the header field;
- the field containing the payload (the data);
- the end of frame field (trailer).

Only the first few bits (1–5, called the 'leading indicator') at the start of the header have different meanings in the static and dynamic segments.

> **COMMENT**
>
> Figure 6.16, which represents the frame, shows only the **bits corresponding to the logical data forming the useful content of the transmitted message** (and not those which are physically carried on the medium); that is:
>
> Header $5 + 11 + 7 + 11 + 6 = 40$ bits or 5 bytes
>
> Payload   from 0 to 127 words of 2 bytes, that is 0 to 254 bytes, that is 0 to 2032 bits
>
> CRC   24 bits or 3 bytes.
>
> Summarising, a communication frame includes between:
>
> $5 + (0 \text{ to } 254) + 3 = $ from 8 to 262 bytes
>
> $40 + (0 \text{ to } 2032) + 24 = 64$ to 2096 bits with logical meanings.

As we will show later, it will then be necessary to add a certain number of **electrical bits**, the purpose of which is to secure the transport of the logical data, in order to form the real electrical signal which is physically present on the network.

Let us now look at the content of these frames field by field.

6.5.2.1   Header Field

The *logical content* (with a logical meaning, in the sense of logical binary data) of the frame begins with the header field. This consists of several parts:

- the *leading indicator*, coded in 5 bits;
- the *slot ID*, coded in 11 bits;
- the *payload length*, coded in 7 bits;
- the *header CRC*, coded in 11 bits;
- the *cycle count*, coded in 6 bits.

So there are 40 bits in total, which will then be transmitted in the form of 5 bytes, the meaning of which we are now going to examine (see Figure 6.18).

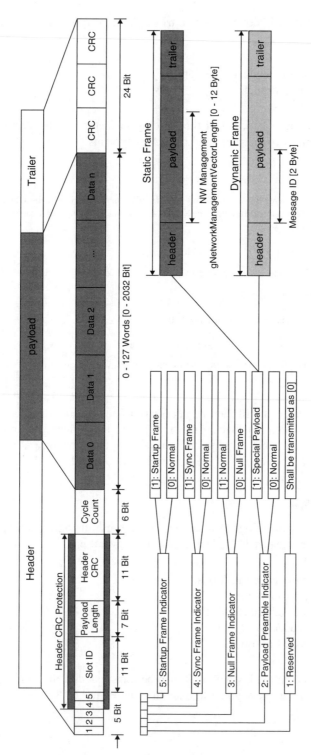

Figure 6.18 FlexRay communication frames: constituent parts

6.5.2.2   Leading Indicator

The first 5 bits form the 'leading indicator'. Their meaning is as follows (in order of appearance):

- **Bit 1:** reserved for future use; the value as of now is 1.
- **Bit 2:** preamble indicator of the content of the payload:
  - '0' indicates classic content of the payload.
  - '1' indicates special content of the payload. Two options are then possible, depending on whether the frame is output in the static segment or dynamic segment:
    * **In the static segment:** The first bytes (0–12) of the payload can signal the presence of a Network Management vector in it. During the slots of the static segments, the communication controller (CC) collects all the available Network Management vectors. At the end of each communication cycle, the CC returns to the host a resulting vector which is obtained in the form of a logical OR of all the vectors of the cycle.
    * **In the dynamic segment:** The first two bytes of the payload contain a supplementary message ID. This can be used, in the receiver, in a receiving node to separate the content of the transmitted frame.
- **Bit 3:** null frame indicator (if a network node outputs a null frame, it means to the network that this node has nothing to report since the previous cycle).
- **Bit 4:** synchronisation frame indicator:
  - '0' indicates transmission of a normal frame.
  - '1' indicates transmission of a synchronisation frame, which must be used to synchronise the clocks of the various nodes which participate in a synchronisation sequence.
- **Bit 5:** start of frame indicator.

6.5.2.3   Slot/Frame Identifier (Frame ID)

Next is the value of the ID of the transmitted frame, the 'frame ID'. This frame ID is coded in 11 bits, giving values from 1 to 2047, the value 0 being considered illegal. As we will see later, its value is used to define the position of the slot in which the frame under consideration is transmitted in the static segment, but also in the dynamic segment.

Additionally, two controllers are not permitted to transmit frames with the same ID value on the same communication channel.

6.5.2.4   Payload Length

The 7 bits (that is, a maximum of 128 values) of this field, called data length coding (DLC), indicate the value – divided by 2 because the transported words include 2 bytes = 16 bits – of the number of words transmitted in the payload segment (maximum 254); that is from 0 to 123. Values above 123 are treated as errors (see below remarks about CRC end of frame).

6.5.2.5 Header CRC

The purpose of the CRC of the header segment is to protect the set formed by bits 4 and 5 of the leading indicator, those of the ID of the frame currently being transmitted and those of the payload length. This enables a node which receives a frame to verify instantaneously the correspondence which must exist between the current slot number and the ID received from the transmitting node, and not to begin work on data which would be unusable.

This CRC is calculated on the fly, online, by the host controller which produces the communication, and its value is, of course, verified in the same way by all the controllers that receive it.

6.5.2.6 Communication Cycle Counter

The cycle counter field, which is coded in 6 bits (so has 64 possible values), indicates the number (from 0 to 63) of the communication cycle in the course of transmission. It can also be used as a 'continuity index' of the communication. The controller which transmits the frame increments it automatically, and its value must be identical for all frames which are transmitted in one communication cycle. Because it is coded in 6 bits, it cannot be incremented indefinitely and is thus periodically recurrent. As we will show later, the cycle number can be used or can assist in multiplexing the frames which a node outputs over time.

6.5.2.7 Payload Field

The field which is dedicated to transporting useful data is the payload field. It provides the possibility of including from 0 to 127 words of 16 bits; that is, from 0 to 2032 bits.

It should be noted that in the payload, it is possible to include 'message IDs' (coded in 0 or 16 bits) (if necessary, see bit 2 of the leading indicator on this subject).

6.5.2.8 CRC of End of Frame

Finally, the frame of logical data ends with a CRC field of 24 bits (3 bytes). Its purpose is to protect the whole of the transmitted frame, with a Hamming distance of 6 for data 248 bytes long or 4 for data above this value (up to 254 bytes long).

This CRC is also calculated on the fly, online, by the host controller which produces the communication, and its value is, of course, verified in the same way by the controllers that receive it.

6.5.3 Encapsulation and Coding of Frames of Logical Data in Slots and Minislots

Now that we have described the aspect of the logical data to be transported, it is necessary to think, on the one hand, about how to transport it on a physical layer where there are some concerns in terms of medium and components for linking to the medium (drivers) and other repeaters, active stars, and so on, and on the other hand, about how to tag them easily as they arrive, with a view to decoding them.

It is therefore necessary to encapsulate the logical data which is described above, and transported during a frame at the time of a static slot or dynamic minislot with the help of protection and precautions for use for transporting data. To do this, let us go back a little to the physics concerning electronics and the propagation of the signal on an electric line, and examine the problems which underline coding and decoding the frames which will be output and received.

6.5.3.1 Bits

In the case of FlexRay, for numerous reasons which will be explained and described in Chapter 8, which is specifically about the physical layer of the protocol, the chosen bit coding is of no return to zero (NRZ) type. Its physical representation (electrical or optical) will also be described in the same chapter.

6.5.3.2 Bytes

Each of the fields which are mentioned in the preceding paragraphs (header, and so on), and which make up the logical content of the frames, are subdivided into bytes, which are then encapsulated in bytes of 10 bits by adding a START bit and a STOP bit. In particular, Figure 6.19 shows how the first $5 + 11$ bits of the 40 bits of the header are repackaged in bytes, and so on.

6.5.3.3 BSS, FSS, FES, Action Point, and so on

Here's another thing! Frame start sequence (FSS), byte start sequence (BSS), action point (AP), frame end sequence (FES), and so on.

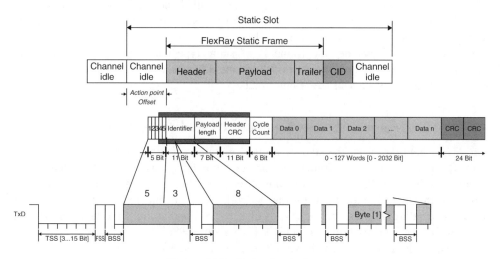

Figure 6.19 Packaging in bytes

BSS (Byte Start Sequence)

Logical bytes are transmitted after framing them carefully at the start with a START bit (at '0') and at the end with a STOP bit (at '1'), thus forming 10-bit bytes of NRZ 8N1 type. As Figure 6.19 indicates, it is true that an 8-bit byte framed by a START and a STOP and followed immediately by a new START gives a unique, easily recognisable aspect to the boundary between two successive bytes. This 'STOP–START (1–0)' symbol/sequence is called a 'byte start sequence' in FlexRay. It is therefore present before every start of a byte with a logical meaning.

FSS (Frame Start Sequence)

To signal the arrival of the first of the bytes of the frame (so before the first BSS appears), FlexRay has included a signal to indicate the start of the frame, called the 'frame start sequence', the rising leading edge of which is, in principle, intended to signal the presence and the start of sending a frame.

This presence gives a very specific appearance to the binary symbol formed by the set of FSS + BSS ('1–1–0'). It can very easily be tagged, and appears only at the start of a frame which is sent in a slot.

In principle, the encapsulation of the logical frame in that of transport should be finished, but alas, the medium and topology which have been adopted for the network still hold plenty of surprises for us. These are the reasons which lead us to introduce a new element, the TSS! What a strange name, full of mystery, isn't it?

6.5.4   ... for Frames which are Transported during Static and Dynamic Segments

To define our context more precisely, let us go back for a few moments to an earlier part of our story.

After a brief instant during which the transmission channel is unoccupied ('channel idle', which we will explain later), at an instant called the 'action point', which was initially predefined by the designer, the node that wishes to communicate activates the start of transmission of data on the network.

Independently of whether the frame is of static type (static slot) or dynamic type (dynamic minislot), to signal the start of the frame it has been agreed that first a sequence called the 'TSS' will be sent (see Figure 6.20).

6.5.4.1   TSS (Transmission Start Sequence), Truncation, and so on

The 'TSS' field is sent at the start of frames of static segments, dynamic segments and symbol segments, and is transmitted by the corresponding transmitter in the slot/minislot under consideration. Its purpose is to initialise the start of a transmission sequence. The length of this field is adjustable by the network designer, and it consists of from 3 to 15 bits depending on uses, the topologies of networks, and so on. Pay attention, because many important things are hidden behind this famous TSS.

In fact, for long reasons which we will explain in detail in Chapter 9, which is about network topologies, it is necessary to take account of all the delays due to line driver

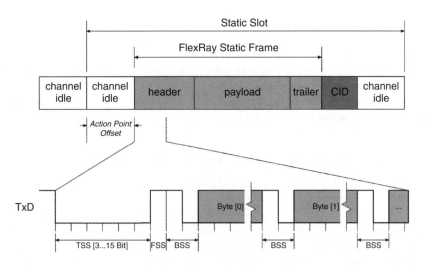

Figure 6.20   Sequence called 'TSS'

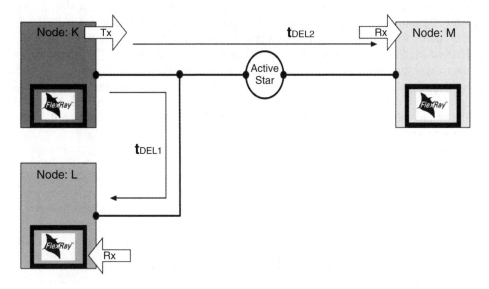

Figure 6.21   Propagation time of signal

components and the effects of distance, paths, obstacles (for example active stars) on the signal path. This obviously causes signal propagation times which differ from node to node (see Figure 6.21).

Despite that, we have to keep global synchronism between the elements of the network, so that each of its participants responds at the right time!

This means that it is necessary to decide on a specific policy, the purpose of which is to compensate for, or ingeniously annihilate, all the possible variations of signal transport time. This is what has been done, and is what is hidden behind the field of bits forming the TSS. Additionally, whatever happens you should know that a receiver on the network requires only a single bit in the low state of the TSS to become aware that a frame is being started somewhere on it. So, sending 3–15 of them at the start seems like waste or madness! Yes, but madness after reflection!

6.5.4.2 Action Point, Truncation and Time Reference Point

Before explaining this madness, let us begin by defining what we will call the 'action point (AP)'.

After the official start of the slot (or minislot) and a period of calm on the network called 'channel idle', the AP is the precise instant at which a node carries out a specific action in accordance with its local time base. In this case, this corresponds to the instant at which the transmitter Tx effectively starts transmission of a frame (see Figure 6.22).

More precisely, if not to be excessively lucid, a receiver node on the network does not have direct knowledge of the instants at which the APs of the static and dynamic slots are produced in the other nodes.

To overcome that and ensure that time is the same throughout the network, a clock synchronisation algorithm (which we will explain in detail in Chapter 14) requires that at each node, when it is in the reception phase, a measurement is carried out of the time difference which exists between the real instant of the arrival of the AP of the static slot of the transmitting node, when the latter sends a synchronisation frame, and the AP of the static slot of the corresponding slot which is imagined or assumed by the receiving node. Thanks to that, the receiving node can deduce the instant of the AP of the transmitters (in a similar philosophy to the 'resynchronisation' of CAN), and compensate for the effect of the propagation delay of the signal.

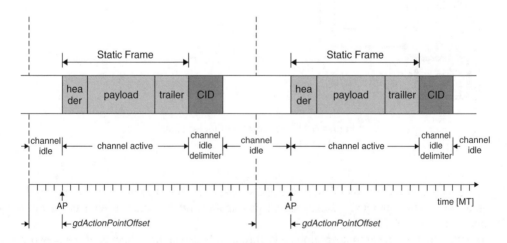

Figure 6.22 Propagation time of signal

Let us examine the concrete case of a transmission from the transmitting node M to the receiving node N:

- First, let us take it as an initial hypothesis that the propagation time/delay always keeps the same value, whatever the rising and falling edges of the signals transmitted by node M and received by node N.
- On the other hand, because of certain effects or devices which may be present on the physical communication medium, it is (highly) possible that the (first) starting edge of the communication frame is delayed for longer than the other edges of the same frame, with the result that the value of TSS seen from the input terminals of the receiver becomes shorter than what is actually output and transmitted. This effect, called 'TSS truncation', has various causes:
  - delays due to the electronic circuits of the line driver stage for putting itself into reception or transmission;
  - delays due to actuating the connection for passing through a coupler in an active star, to know in what direction the exchange from an input x to an output y passes;
  - and so on.
- Obviously, the truncation effect of the TSS sequence, due to all these causes and successions of causes, is cumulative on the TSS values which are transmitted and received between nodes M and N, and reduces the length/duration of TSS. Despite that, if during the time phase reserved for TSS a consecutive number of bits are detected in the low state in a range from 1 to ($gdTSSTransmitter + 1$) bits, a node must accept this signal as a valid TSS.
- Obviously, it is the duty of the system designer/architect (who essentially has the knowledge of the topology and components of the network which he or she wants to make operational) to take account in detail of these facts, and consequently to calibrate, at the level of each node of the network, adequate values to be given to TSS (from 3 to 15 bits).

Figure 6.23 describes and emphasises again the separate effects of, on the one hand, pure signal propagation effects and, on the other hand, the effect of truncation and the meaning of 'TSS truncation'.

As we have just indicated, because of TSS truncation and signal propagation delays, it is impossible to know, easily and precisely, the precise time relationship between the

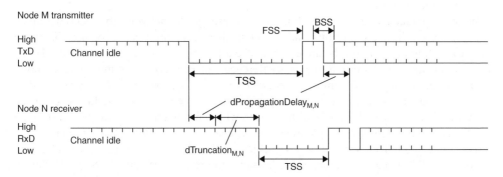

Figure 6.23   TSS truncation

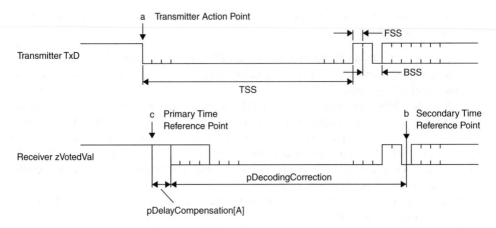

Figure 6.24 Action point

instant when the receiver begins to see the TSS and the instant when the transmitter started to send it. It is then necessary to reference the time for measuring the received frames using an element of the transmitted frame which is not affected by TSS truncation. This is why FSS and BSS, among other things, were also designed!

6.5.4.3 Action Point and Determination of the Action Point

Now that the BSS exists and its form is known and easily recognisable, let us go back to the succession of actions which will lead us to knowing more precisely the instant of the AP. To do this, let's examine Figure 6.24:

- Independently of the rest of the world, in its static slot, the transmitting node starts the transmission of the static frame which it wants to transmit, from the instant corresponding to its own AP *a*, including in the transmission of its message the values of TSS, FSS and the first BSS.
- After propagation, delays and possible truncations, the message is received in a receiving node of the network.
- Whatever incidents have occurred on passing through the network, the receiving node, by sampling the incoming bits and decoding their structure, can recognise the shape and structure of FSS and the appearance of the two bits forming the first of the BSSs.
- From now and by definition, the time reference called 'secondary time reference point (TRP) timestamp – timestamp *zSecondaryTRP*' – is taken as the instant (measured in local microticks) at which the sampling point of the second bit of the first BSS of the frame of the message (point *b* of the figure) which forms the potential start of a frame (that is, the first HIGH to LOW transition detected after a valid TSS) is produced.
- This time reference 'secondary TRP timestamp' will now be used to calculate the primary TRP timestamp (point *c* of the same figure), which represents the instant at which the local node should have seen the start of the transmitted TSS if the value of TSS had not been affected by a truncation effect and propagation delays.

- The primary point TRP (timestamp *zPrimaryTRP*) is then calculated (in local microticks) from the secondary TRP, by subtracting a fixed offset *pDecoding-Correction* (to correct certain delays due to the decoding process) and a delay compensation term *pDelayCompensation* which is due to the effects of propagation delay of the signal on the network. The time difference between *zPrimaryTRP* and *zSecondaryTRP* is therefore the sum of the node parameters *pDecodingCorrection* and *pDelayCompensation*. It should be noted that this does not quite represent the real situation, but instead indicates the instant which the timestamp represents.
- The primary TRP timestamp (point *c* of the figure) is (or will be in Chapter 14) used as the 'observed arrival time' of the frame by the clock synchronisation algorithm.
- As we will show, this algorithm uses the measured gap between *zPrimaryTRP* and the arrival time of the expected frame to calculate and compensate for the gaps of the local clock of the node.
- Following this calculation, the decoding process will then supply the output signal 'potential start of frame in A' to the clock synchronisation startup (CSS) on transmission channel A.

We have just completed the encapsulation of the logical data of the frame to be transmitted, from the first bit of the header to the last bit of the CRC. There are only a few microdetails of 'adjustments', and the matter will be closed!

Let's start with the first one.

6.5.4.4   FES (Frame End Sequence)

Now that we know the structure of the encapsulation of communication frames from their openings to the CRC, it is time to close them, if only to prevent drafts. To do that, a symbol consisting of two bits ('0–1') is provided, and as anyone might have guessed, it is called the frame end sequence.

An example is given in Figure 6.25 in the case of a dynamic end of frame.

6.5.4.5   CID and DTS

CID (Channel Idle Delimiter) – Static (and Dynamic) Frames
Outside the frame of logical data and its transport protection CRC, all encapsulated as above, to fill the time between the end of the electrical frame as such (whether it is static or dynamic) and the end of the slot, the structure ends with a field of 11 '1' bits called the 'channel idle delimiter (CID)', the purpose of which is to signal the end of transmission of a frame in the static slot, and to free the medium to leave it idle. Figure 6.26 shows the end of a static slot.

DTS (Dynamic Trailing Sequence) – Dynamic Frame
It should be noted that there is a small problem about transmission of dynamic frames in dynamic segments. In fact, it is necessary to add an element – the dynamic trailing sequence (DTS) – between the end of the CRC and the start of the CID, since when the minislot is opened to become a dynamic slot with a specific, variable duration, one must fall back cleanly on one's feet in relation to the durations (calibrated in microticks) of the minislots, and above all depending on the precise position of the AP of the next minislot.

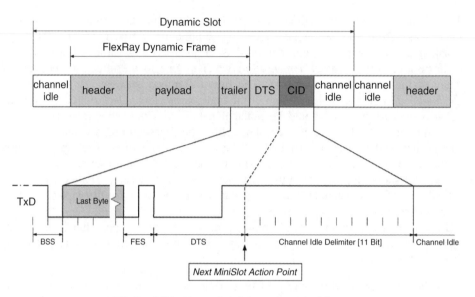

Figure 6.25 Example of dynamic end of frame

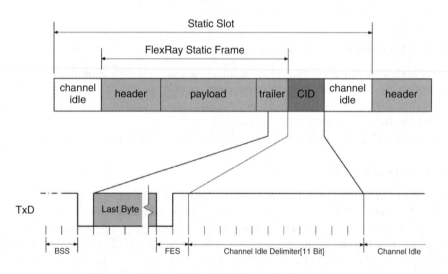

Figure 6.26 End of a static slot

This is the reason that has led to the introduction of this little buffer of variable duration, the DTS, which is found only almost at the end of dynamic frames.

Figure 6.27 shows an example in which a minislot equals five macroticks, and how the value of DTS must be adjusted so that the next output frame is able to start exactly at the instant of the AP, after the known 'channel idle' period.

As a conclusion to these paragraphs, Figure 6.28 gives a summary of the end of a dynamic slot.

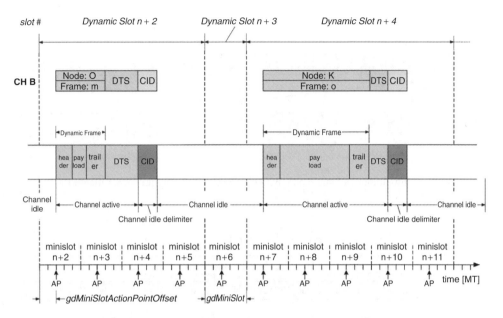

Figure 6.27 Example with a minislot equal to five macroticks

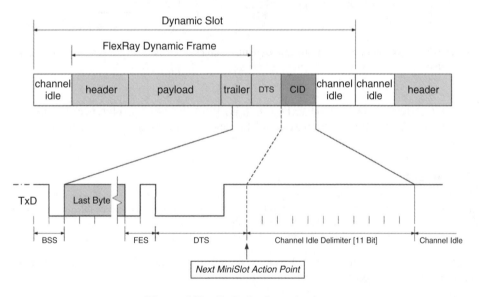

Figure 6.28 End of a dynamic slot

6.5.4.6 Complete Static and Dynamic Frames

To conclude these long paragraphs, Figures 6.29 and 6.30 summarise the configurations of static and dynamic frames after they are encapsulated for transport.

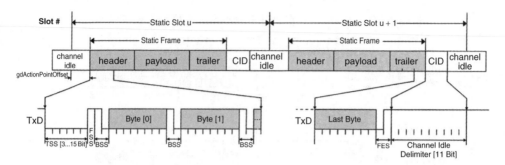

Figure 6.29 Configurations of static and dynamic frames after they are encapsulated for transport

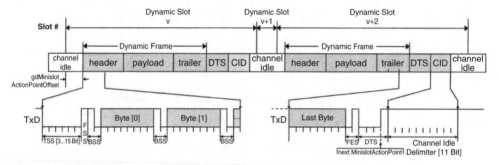

Figure 6.30 Configurations of static and dynamic frames after they are encapsulated for transport (continued)

As a conclusion and summary, Figure 6.31, from the official reference documents of the FlexRay Consortium, summarises all the encapsulations we have mentioned in this chapter.

COMMENT

Like a teacher, we too could have 'encapsulated' our chapter by putting this same figure at the start of it, so that it acts as a guide, but we preferred to present it to you only at the end of the presentation, as a summary, because this is what you have to remember!

Before concluding this section, let us examine a last point: what could be the length of the longest transported frame?

6.5.4.7 Maximum Binary Length of a Static Frame

As we have just shown, to secure the transport of logical data, it has been necessary to add a certain number of **electrical bits** (TSS, FSS, BSS, START bit, STOP bit, FES, DTS, CID, and so on) to the bits corresponding to logical data, to form the real electrical signal which is physically present on the network. Taking account of all these additional

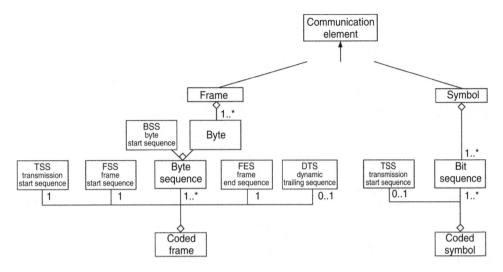

Figure 6.31 Encapsulations

bits, it is easy to calculate the maximum physical length of the longest static frame which can exist on a FlexRay network.

Knowing that the communication frame includes, at maximum:

- $5 + (0-254) + 3 =$ from 8 to 262 bytes with logical meaning,
- that is $40 + (0-2032) + 24 = 64$ to 2096 bits with logical meaning,
- and that

TSS one per frame	Maximum 15 bits
FSS one per frame	1 bit
BSS before each byte	2 bits
FES one per frame	2 bits
CID one per frame	11 bits

- in total, the maximum possible number of electrical bits is 2638.

Figure 6.32 summarises all these points.

At a maximum bit rate of 10 Mbit/s or 1 bit $= 100$ ns, that corresponds to a maximum frame duration of about 0.270 ms.

In conclusion, this shows that in the best case, FlexRay can have data transport efficiency in a static slot of

$$(\text{max\_useful\_bits} = 2096)/(\text{max\_total\_bits} = 2638)$$

giving a transport yield of

$$\text{useful rate/max. rate} = \text{about } 80\%$$

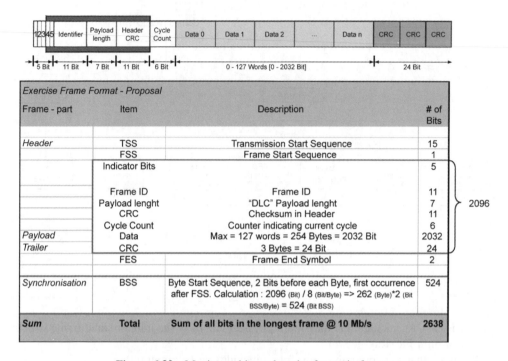

Figure 6.32 Maximum binary length of a static frame

For information, that of CAN was only about 50% in the best case, mainly because very little data (maximum 8 bytes) was transported per frame.

Additionally, the FlexRay specification indicates that the maximum cycle time is limited to 16 ms, which in the case of use of a static segment only would make it possible to transport a maximum number of about 60 of the longest frames (see another example in Figure 6.33). Does this really mean anything? That's another story!

To complete the exploration of the FlexRay communication cycle, let us now go on to examine the last two little segments (in size but not in importance): those of 'SW' and 'NIT'.

6.6 'SW – Symbol Window' Segment

This optional segment is reserved for the SW, which is dedicated for inclusion of the media access test symbol (MTS). The MTS is used to verify that the local bus guardian is functioning properly. The MTS has the same structure as the collision avoidance symbol (CAS). The length of the two symbols is 30 bits low.

To terminate the SW segment, the MTS is followed by a conventional CID (see Figure 6.34).

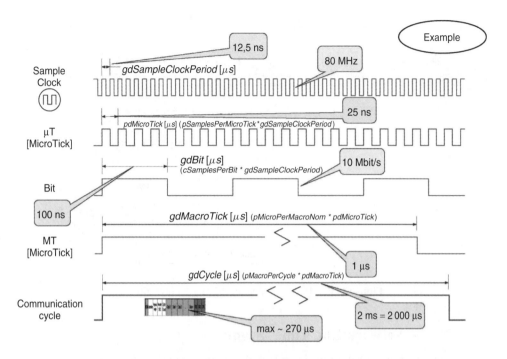

Figure 6.33 Maximum binary length of a static frame

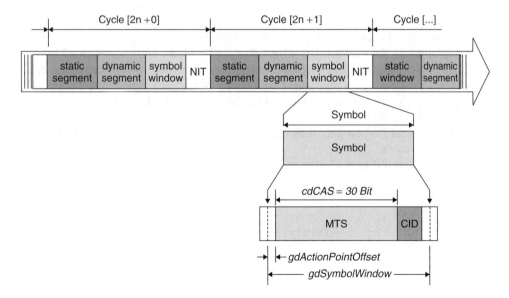

Figure 6.34 The media access test symbol is followed by a conventional CID

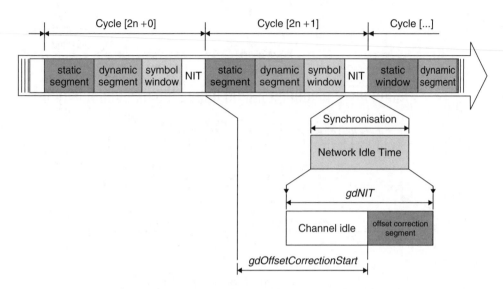

Figure 6.35   Local application of phase corrections

6.7   'NIT – Network Idle Time' Segment

Each FlexRay communication cycle ends with a very special segment, the NIT.

Seen in the oscilloscope, nothing happens on the network, and no traffic occurs on its lines while it lasts. The network is in idle (waiting) mode. That's the surface view – but under the surface, there is great activity, which justifies the presence of this NIT segment, the duration of which must not exceed 767 macroticks. In fact, during the start of NIT, all the nodes of the network take advantage of it to do the necessary calculations for global synchronisation of the network (in offset/phase and rate, as will be described in great detail in Chapter 14), and then, as shown in Figure 6.35, during its last part all the nodes take advantage of it to apply their phase corrections locally. Since these phase corrections affect all the participants in a group (cluster), the duration of NIT is therefore a quantity associated with a cluster.

This is the end of the first part of the guided tour of the FlexRay protocol and a FlexRay communication cycle.

After these plentiful hors d'oeuvres, let us now go on to the numerous main courses.

7

Access to the Physical Layer

To understand properly all the technical, industrial and economic refinements which are included in the possible applications of FlexRay, let us return in detail to the techniques which are used regarding the types of access to the medium, which are one of the fundamental, specific points of this protocol.

7.1 Definition of Tasks

To begin and as a reminder, it is the duty of every designer to define the tasks of each node of the network (point-to-point links, centralised tasks, and so on, most often functioning in distributed tasks), and to define carefully the tasks which must access the network in deterministic or real time manner or with known latency time ... or not.

The designer then has, conceptually:

- on the one hand, in the communication cycle, the ability to choose (once and for all) the relative distribution between the durations of the static and dynamic segments relative to each other (see Figure 7.1), while remembering that in principle there is neither encroachment nor interference between these two segments, and that completely different data can be sent on the two channels during the same time slot, and that different nodes can use the same time slots on different channels;
- on the other hand, two possibilities for access to the network (see the appendices to Part B):
  - either via the static segment, for 'quasi real time' tasks or those with a predetermined, known latency time, making it possible, if required, to use multiple time slots per node, so as to be able to access the network during the same communication cycle;
  - or via the dynamic segment, to serve tasks which are triggered asynchronously by events ('spontaneous, event-triggered'), subject to hierarchisation of access (often wrongly called arbitration) and have a bandwidth which can be adapted dynamically in operation and according to the operational requirements of the whole.

Let us use the above-mentioned figure to explain the operation and possibilities of this very special configuration. This figure shows the whole of the two communication segments, static and dynamic, with a separation which can be determined as desired by the system designer.

FlexRay and its Applications: Real Time Multiplexed Network, First Edition. Dominique Paret.
© 2012 John Wiley & Sons, Ltd. Published 2012 by John Wiley & Sons, Ltd.

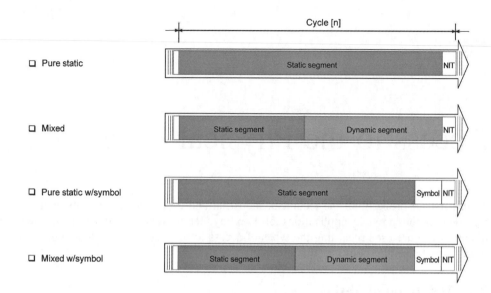

Figure 7.1   Relative distribution among durations

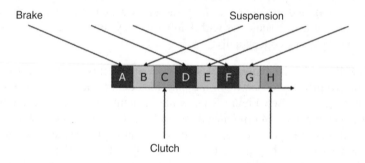

Figure 7.2   Access (time slots)

The choice is made according to what the network designer himself or herself wants. It makes it possible to complete the allocation of bandwidth (in terms of megabits per second) according to what is wanted for the imagined system to operate, having divided the payload into two segments of sufficient dimensions.

Once that has been done, it is necessary to assign to each of the network participants which need access in quasi real time, within the static segment, time slots which will belong to those participants. We have shown this in Figure 7.2, in which, for instance, we see that application A (braking function) can communicate during the first time slot, application B during the second time slot and application E (clutch control) during the fifth time slot, and so on.

Up to now, it's all simple! But there is an enormous hidden face, with a double release, behind the mere technical presentation and interpretation of this simple little figure. Let's look at that more closely.

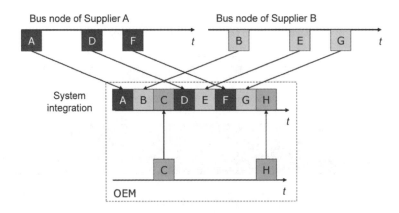

Figure 7.3 Allocation

In fact, because of the mere principle of functional assignment of time slots and the fact that it is certain that time slots do not overlap, this enables the system integrator (for example the car manufacturer or industrial group) to choose as it wishes – for reasons of skill, cost, testability and so on – the same single equipment manufacturer/supplier X for applications A, E, H, and so on, and another, Y, for the other applications, as shown by the allocation in Figure 7.3.

Apart from the fact (as some malicious people might think ☺!) of being quite free to divide and rule, this then enables the system manager to integrate all the modules much more quickly than in CAN, since all the data which serve the applications are strictly separated, with no concept of arbitration or random latency times of tasks. The integration and test time, and of course their costs, are thus enormously reduced, and the flexibility of development is greatly increased, implementation of variants is made easier and the choice of competing industrial partners to implement the system is greater. All that represents an undeniable advantage of FlexRay compared with other protocols.

Once the allocation of static segment and dynamic segment is defined in the communication cycle, the designer must define the 'slicing' of each of them, to determine how many slots will make up the static segment and how many minislots will make up the dynamic segment. In the case of using two FlexRay transmission channels, it should be noted that firstly this time slicing must be identical on both and secondly the FlexRay specification also indicates limits greater than these values.

A first example is given in Figure 7.4.

At this stage, the total number of slots (static and dynamic) is entirely determined, and it is easy to assign to each of them a number from 1 to xxx within a communication cycle, at least on paper, one after the other.

COMMENT

At the end of each chapter, we will present a concrete application example, which makes it possible to understand how one can define the number and format of slots and minislots and communication frames.

Where we are now, in the static segment, being in a purely deterministic architecture for access to the network, means that as of now we know which frame of which

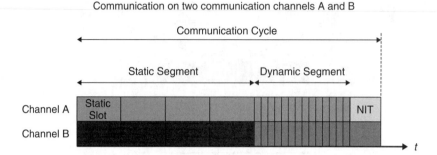

Figure 7.4   Example

communication cycle of which node will be sent in a given static slot. This means that there is a direct correlation between the name of the node, the slot number and the desired function.

To set that in concrete and avoid any error in assignment of slots, all that is now necessary is to provide a direct correlation between the cycle number and the number of the desired slot, and to associate with them the frame that one wishes to send, to ensure that the right frame is sent in the right slot. For this purpose, the 'frame identifier' (ID) in the header of the content of the frame to be sent, and the 'arbitration grid' mechanism, which we are now going to describe, have been introduced.

7.2   Execution of the Communication Cycle

Apart from what happens during the network startup phase, the repetitive communication cycles are numbered repetitively, from 0 to *cCycleCountMax* (=63), and are executed periodically, with a period of which the duration is formed by a finite, constant number of macroticks.

Arbitration (actually, being more purist, 'hierarchisation' of access to the network – see below) within static and dynamic segments is based solely on the frame ID function, which is assigned to the nodes in the cluster (group) of each channel, and on a counting scheme which is supplied by the numbered transmission slots.

During the static segment, actual access to the network is carried out using a two-level sieve, the first level being the frame ID and the second the 'arbitration grid'. Let us first examine what the frame ID contains.

7.3   Frame ID (11 Bits)

As defined by the FlexRay protocol, the frame ID indicates and defines the slot in which the frame under consideration must and will be transmitted. Also, a frame ID can be used only once per channel and per communication cycle. Because the value of the frame ID is coded in 11 bits, the range of values which it is possible to assign to the frame ID is between 1 and 2047, since the value frame ID = 0 is invalid.

The first consequence of the above paragraph is that each frame to be transmitted in a cluster has a frame ID assigned to the cluster.

To plan the transmission which it must provide in the course of the communication cycle, each node of the network is obliged to maintain within it a variable called *vSlot-Counter*, which indicates at all times the state of the slot counter, one for channel A and a second for channel B. These two counters are reset to 1 at each start of each of the communication cycles and incremented by one at each end of slot, whether they are of static or dynamic type.

The transmitted frame ID is determined by the value of the slot counter *vSlotCounter (Ch)* at the instant of transmission. In the absence of error, the value of *vSlotCounter (Ch)* can never equal zero when a slot is available for transmission. Received frames with the number zero are therefore always identified as frames with errors, since the slot ID must be wrong because there is no slot with an ID equal to zero.

Once the network designer has fixed the duration of the communication cycle, the position of the separation between the static and dynamic segments and the number of slots and minislots in their respective segments, the value of the frame ID determines, on the one hand, the transmission slot and, on the other hand, as a consequence, in which segment and at what instant within the segment the associated frame will be sent. By definition, therefore, the range of frame ID is between 1 and *cSlotIDMax*.

The node transmits its frame ID most significant bit first, and then the other bits of it are transmitted in descending order of significance.

7.4 Arbitration Grid Level

In the time hierarchy of a network which functions under FlexRay, the level called 'arbitration grid level' is immediately below that of Global Time, which contains the arbitration grid, which forms the backbone of the arbitration principle (see comment) for access to the FlexRay medium.

The rules of the arbitration grid differ depending on whether the frame that the node wishes to transmit is intended to go into the static segment or the dynamic segment of the communication cycle:

- in the static segment, the arbitration rule is based on the construction of time intervals called static slots;
- in the dynamic segment, the arbitration rule is based on the construction of time intervals called minislots.

IMPORTANT COMMENT ON THE TERM 'ARBITRATION'

In the course of the various talks, courses, lectures, presentations and technical training about the FlexRay protocol which we have done, we have realised that the meaning and understanding of the term 'arbitration' often leads to confusion, depending on the origin of the knowledge that each person has. Apart from what it strictly means according to the dictionary, let us briefly consider two important meanings which are generally assigned to this word.

Dictionary definition: Action of arbitrating: Arbitrating in a rugby match. Arbitrating in a conflict (syn. conciliation, mediation). Decision made by an arbiter: the arbitration is unfavourable to him (syn. judgement, verdict) (definitions from Larousse dictionary, translated; 'referee' is 'arbitre' in French).

Without wanting to play with words, if you say 'arbitration' there must be something to arbitrate; that is, disputing claimants and, depending on how aggressive they are, a fight or not.

Let's take a specific example of clear aggression. This is the case of the CAN protocol (or I2C), in which, when the medium is considered to be free by the participants which may want to access the medium, with no other form of process, after a predetermined time, all the participants of the network try to take it, strictly at the same time. It is therefore necessary to arbitrate in real time, on the fly, almost bitwise, on everything that occurs to manage access to the medium, and to go on to real, muscular arbitration (usually using strong-arm tactics with the bits with physically dominant representation and against the poor little recessive bits, which are completely crushed).

Regarding FlexRay, this happens in a non-aggressive manner, since the word 'arbitration' is used here in a more civilised, polished sense, with no latent fighting; that is, within the philosophical framework of taking care to respect politely the priorities to which all have rights. This technique resolves a kind of arbitration between protagonists, without managing conflicts. This is more to do with respecting a predetermined, known hierarchisation of the protagonists.

7.4.1 Basic Concepts

The two segments, static (ST) and dynamic (SD), which are defined in a FlexRay communication cycle each consist of a number of slots:

- The static segment consists of static slots, which are characterised by the parameter *gNumberOfStaticSlots*, which fixes the number of slots in the static segment, and the parameter *gdStaticSlot*, which fixes the size of a slot (all static slots have the same size).
- The dynamic segment consists of a number (*gNumberOfMinislots*) of minislots, each of duration *gdMinislot*.

7.4.2 Policy for Access to the Medium

A FlexRay communication cycle allows two modes for access to the medium:

- GTDMA (for generalised or global time division multiple access) for the static segment, making it possible to allocate several slots to the same node in the same cycle;
- FTDMA (for flexible time division multiple access) for the dynamic segment.

As we have indicated, each node maintains within it a cycle counter *vCycleCounter*, the purpose of which is to provide information on the value of the current communication cycle. Also, each node maintains a slot counter, which is initialised at each start of a communication cycle and incremented at the end of each slot.

Unlike static slots, dynamic slots can be of different sizes. In fact, if a node actually has data to transmit in its associated slot, the size of the dynamic slot will be equal to the size of the relevant message; otherwise the slot has a minimum size equal to *gdMinislot*. In a static or dynamic slot, a single node is authorised to transmit. This is the node which holds the message ID (frame ID) which equals the value of the slot counter. The ID of a frame consists of 11 bits, making it possible to define a maximum of 2^{11} different IDs {1 ... 2047}. The allocation of IDs to nodes is static and decided offline. Each node which is going to transmit messages has one or more static and/or dynamic slots which are associated with it. Non-collision at the level of a slot is resolved by allocating a slot to at most one node. Thus, two different nodes cannot transmit at the same instant; that is, in the same slot.

In a FlexRay network, each node consists mainly of a CPU/ECU (electronic control unit) and a communication controller (CC). They are interconnected by a controller-host interface (CHI). The controller implements the services which are defined by the FlexRay protocol, and the CHI manages the flow of data and control between the CPU and the CC. At each node, the CHI reserves buffers where the CPU can write messages to be transmitted. At the start of each communication cycle, the cycle counter is incremented and the CC reads the buffers to prepare the frames to be transmitted in the current cycle.

7.5   Conditions of Transmission and Access to the Medium during the Static Segment

Transmission of a static message, that is one which is transmitted in the static segment, is based on a table which is generated offline, and which, as well as the slot number, defines two additional parameters: the message frequency and its offset.

These two parameters make it possible to define unambiguously the precise instant at which the message must be transmitted.

EXAMPLE

A message with frequency 5 and offset 2 means that the message is transmitted in the second communication cycle and every five cycles.

The static segment is thus quite suitable for communicating messages of time-triggered type. In other words, a message is transmitted each time its slot/window arrives, even if the data have not been updated.

In the static segment, even if a node has no data to send in a given slot of a given cycle, this slot will always be the same size, but will have no data in the payload field. The size of the static slot (the same for all slots) is a global parameter of the network, and can and must be fixed according to the size of the longest message which the user has to transmit. However, this approach is very limiting. In fact, if it is considered that the FlexRay specifications fix the maximum possible duration of the communication cycle at 16 ms, it would be preferable to consider the possibility of dividing excessively long messages into several frames, and grouping several short messages, to equalise the traffic. Without great difficulty, one can find methods which make it possible to optimise the length of the communication cycle (*gdCycle*) and the number of static slots (*gNumberOfStaticSlots*), to maximise the number of unused slots and thus make future developments of the system possible. In general, these techniques model the problem of arranging the messages in the static segment by reducing it to a problem in integers, with multiple choices and a non-linear cost function. However, if a mechanism for managing the packing of messages is implemented, it is necessary to take account of the additional communication costs (see the appendices to Part B).

7.6   Conditions of Transmission and Access to the Medium during the Dynamic Segment

Let us now explain in detail the structure of access to the network during the dynamic segment of the communication cycle.

Transmission of a message in the dynamic segment is totally different, and is based on the minislotting mechanism. As we are about to show, this segment is more appropriate for communication of messages of event-triggered type, where a message is transmitted each time there is an update in the buffer of the node which is the source of the message.

Once the dynamic segment has been decomposed into a chosen number of minislots, this segment simply waits to be filled with data from the various nodes. Also, the network designer has (or should have) assigned values of frame ID (linked to the node, dynamic slot and channel) to the messages which will be carried during the dynamic segment.

If a node actually has a message to send in this segment, and can do it successfully, the corresponding dynamic slot will have the size of the relevant message; otherwise, it will keep the size of a minislot. The previous sentence is ambiguous, and indicates that the 'flexibility' which is provided by the dynamic segment is nevertheless badly controlled. Let us explain that, while giving some additional details.

By definition, during this segment, access to the medium is based on the values of frame ID, which define 'hierarchies' for access to the medium by frames. The access rule is that the lower the binary value of the ID, the higher the priority of the frame. In principle, it couldn't be simpler! Except that ... one of the principal problems of access to the medium in this segment is also the fact that a frame which is ready to be sent by a node can be delayed by one or more communication cycles before accessing the network medium and actually being transmitted on it.

In fact, the dynamic segment consists of a finite number of minislots. Each time a message is transmitted in this segment, the minislot counter is shifted by the length of this message. Thus, when the slot counter reaches the value corresponding to the ID of the message which is ready to be transmitted, the CC checks whether there are still enough available minislots (counted in microticks) to make it possible actually to send the frame of this frame ID. This operation is carried out continuously by comparing, at every instant, the value of the minislot counter with the parameter *pLatestTx* (a global parameter at the level of a node, and equal to the size of the longest message which this node will have to transmit).

To illustrate the consequences of what we have just described, let us examine the three different application cases which can occur:

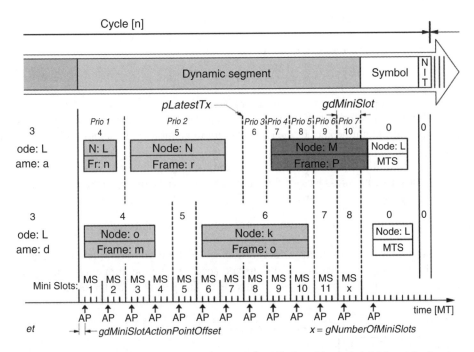

Figure 7.5   The sequences of minislots restart while keeping their initial numbering

1. Because of its application task, at the time of a particular communication cycle, the node under consideration has no 'event-triggered', non-real time message to send on the network.

   In this case, the minislot which is dedicated to its potential messages (via the value of its frame ID linked to the node, dynamic slot) is not occupied, and therefore, since it has not occupied its hierarchical rank (its minislot) in this communication cycle, it leaves to other potential users of the dynamic segment of this cycle the opportunity to use the remaining available bandwidth (the subsequent minislots) of the dynamic segment in this cycle.

2. Because of its application task, a node wishes to transmit during the minislot which is assigned to it in the dynamic segment.

   When the number (ID) of the relevant minislot of the dynamic segment presents itself, the node then jumps on the moving train, and its minislot is used for the dynamic frame which it wishes to transmit.

   Consequently, when a node has access to one of its minislots, the duration of this minislot, which keeps its number, is increased depending on the content of the message, and therefore changes its name from minislot to 'dynamic slot'. When the node has finished transmitting, the sequences of minislots restart, while keeping their initial numbering (see Figure 7.5).

   That was the simple case! In the original text of FlexRay: 'If Minislot is earlier than pLatestTX: "Message cannot be sent"'.

3. The problem can become a little more complicated when a node wishes to send a long message and/or one of low priority.

Let us take as an example the scenario described in Figure 7.6.

It is assumed that at the start of the communication cycle which is in the course of starting, all the messages are ready to be transmitted:

$$\{m_1(pLatestTx_{N1} = 9),\ m_2(pLatestTx_{N2} = 6)\ \text{and}\ m_3(pLatestTx_{N1} = 9)\}$$

where m_1, being the message with the highest priority, is sent in the first cycle.

When it has been transmitted, the minislot counter is at 8 ($>pLatestTX_{N2} = 6$). Message m_2 can therefore not be sent in the current communication cycle, although it has higher priority than m_3.

Consequently, the slot counter in the dynamic segment may not reach its maximum value, thus forcing messages of lower priority to wait for the next communication cycle to try to access the network.

■ Again, in the original text: 'If Minislot is later than pLatestTX: "Message cannot be sent"'.

The purpose of the paragraphs which follow is to show you some complex and very characteristic examples of access conditions and access to the medium during the dynamic segment of a communication cycle. Appendix B2 of this part will add the finishing touches to the whole.

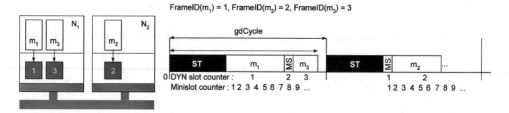

Figure 7.6   Scenario for transmission in the dynamic segment

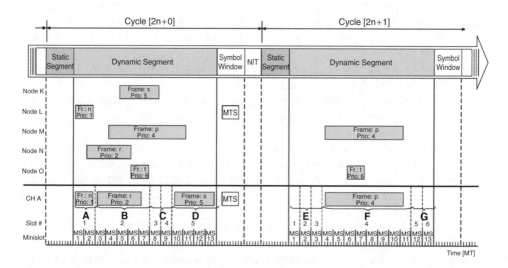

Figure 7.7   Example 1

7.6.1   *Access to the Medium during the Dynamic Segment – Example 1*

Let's look carefully at the example shown in Figure 7.7.

At the time of cycle $2n + 0$, five candidates for communication are ready to jump into their respective minislots – if they can. Let's look at this problem, not entirely seriously.

Cycle number	Frame name	Hierarchy level	Comments and explanations
$2n + 0$	n	1	Frame n of hierarchy '1' is small and can pass. It occupies minislots 1 and 2. Interrogation of hierarchy '2' is therefore shifted temporarily to the start of minislot '3'
	r	2	Frame r of hierarchy '2' can also pass, because of its size. It occupies minislots 3, 4, 5, 6, 7 and also shifts the subsequent minislots
	–	3	No candidate at hierarchy '3'! Is there anyone at hierarchy '4'?
	p	4	Yes – but the size of frame p of hierarchy '4' is too great to enter the remaining time (counted in macroticks) of the dynamic segment. Its access to the network is therefore refused. No chance for it! Does it want to try another chance in the next cycle? It must decide
	s	5	Is there anyone at hierarchy '5'? Yes! Can the size of frame s enter the remaining time? Yes, so it's gone!
	t	6	Sad for '6'! The whole segment is full. Same comment as for hierarchy '4'
$2n + 1$	–	1	No-one at hierarchy '1'! Is there anyone at hierarchy '2'?
	–	2	No-one at hierarchy '2'! Is there anyone at hierarchy '3'?
	–	3	No-one at hierarchy '3'! Is there anyone at hierarchy '4'?
	p	4	Yes! The one in the previous cycle has tried its luck again, and now, even allowing for its size, it can manage to pass
	–	5	Is there anyone at hierarchy '5'? No! Is there anyone at hierarchy '6'?
	t	6	Yes – but rather large! Its access to the network is therefore refused. Hard luck! Does it want to try its luck again in the next cycle? It must decide
			etc.

7.6.2 Particular, Difficult Choice of Hierarchy of Frame ID – Example 2

The definition of hierarchies and of the relationship between hierarchy and message length is often tricky to state and use. Let's give a (bad?) application example.

Let's take the example of implementing the toggle switch by which a window in a vehicle can be both lowered and raised. Merely for teaching purposes, we will assume that this system does not include an anti-pinch system – although that is mandatory!

It may be considered that the action of lowering a window is occasional, and hardly real time, and thus that the hierarchical level of this function is also quite low. We have decided to assign a frame ID at the end of the dynamic segment, in order to leave other frame IDs with higher priority, which may mean waiting for a few communication cycles to pass the message. But lowering a window takes longer than several multiples of 16 ms! In principle, the raising action is at the same level. But if a child happens to put a hand or head through the window while it is being raised, and you want to stop or lower the window immediately, what do you do, what happens? It's a different game! Despite the occasional nature of the action, a higher hierarchical level is required for safety.

As we said before, this example is completely unrealistic, and was only presented to you to illustrate simple reasons for teaching purposes. It is particularly unrealistic because firstly all electrical window regulating systems have built-in safety with anti-pinch devices, and secondly because up to now no-one has thought of connecting a window regulator to a FlexRay network.

7.7 Similarity of the Use of the Dynamic Segment to the Network Access of the CAN Protocol

The fact that the unique number of the frame – the frame ID – defines the number of the minislot in which the frame can be transmitted (to output the frame, there is a relationship between the minislot number and the ID) means that the transmission hierarchy in the dynamic segment functions – seen at a distance – quite similarly to that of CAN. In fact, apart from 'aggressive' management of conflicts, it is the ID number that presides over the sending hierarchy (which could be called arbitration over taking the medium) on the medium. The lower the frame ID number, the more the first minislots will be occupied, and the more chance there will be to transmit in the dynamic segment – as a function of the possible time for which it will remain in the dynamic segment. There is not, and cannot be, in the strict sense a phase of bit by bit arbitration of two frames that start at the same time, but if two nodes want to transmit at the same time, they will only access the medium as a function of the values of their IDs – again, as a function of the possible remaining time in the dynamic segment. Summarising:

- Access conditions:
  - according to the hierarchical order which was established offline;
  - the duration of the message is compatible with the remaining time in the dynamic segment;
  - the possibility of communicating in the next cycle, on the same access conditions as above.

- For a long time, not entirely seriously, we have, like a teacher, tried to establish the differences of access to the medium during the dynamic segment between 'fighting' in CAN and 'fairness' in FlexRay.
  - **Fighting** In CAN, access to the medium is done with bit by bit arbitration, and the dominant bits crush the recessive ones. This is therefore a true fight to gain access to the network.
  - **Fairness** In FlexRay, hierarchies are established offline, by meeting in an office and deciding on the right precedence. For instance, when the bus arrives at the bus stop (corresponding to the instant of the start of the dynamic segment), first we allow grandmothers to get on, then pregnant women, then grandfathers, then women, then men and finally girls and boys. There can be no conflict if everyone applies the established rule.

All these parameters and the broad range of their possibilities obviously open up a large number of probabilistic conditions for access to the network. This is certainly one of the trickiest points to resolve in FlexRay applications. To help you to get through the maze and try to solve this problem, we refer you to Appendix B2.

7.8   Some Additions in the Case of FlexRay Being Used with Two Channels

To complete this chapter, let us consider the FlexRay applications operating with two communication channels, and their consequences for access to the medium during the respective segments of the two channels.

Everywhere, within the dynamic segment, so as to continue to program/plan transmissions, each node continues to operate the two slot counters, one for each channel. Whereas the slot counters for channels A and B are incremented simultaneously while the static segment lasts, they can be incremented separately and independently, in accordance with the dynamic arbitration scheme, in the dynamic segment.

Figure 7.8 emphasises the scheme for access to the medium within the dynamic segment.

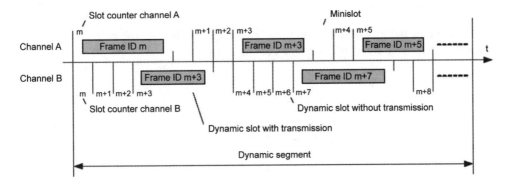

Figure 7.8   Access to the medium in the dynamic segment

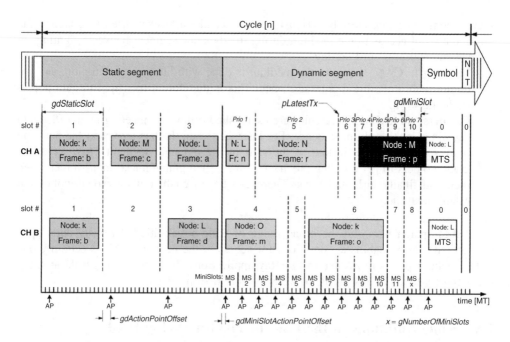

Figure 7.9   Access to media A and B

As Figure 7.9 shows, although the two channels use strictly the same arbitration grid based on the same minislots, access to media A and B and communication in the dynamic segment do not necessarily occur simultaneously.

The number of minislots *gNumberOfMinislots* is a global constant for a given cluster.

The node does the housekeeping work of maintaining its slot counters on the basis on one per channel. At the end of each dynamic slot, the node increments the slot counter *vSlotCounter* by one. This takes place until either:

1. the relevant channel slot counter has reached the value *cSlotIDMax; or*
2. the dynamic segment has reached the minislot *gNumberOfMinislots;* that is, the end of the dynamic segment.

When one of these conditions has been reached, the node sets the corresponding slot counter to zero for the rest of the communication cycle.

Appendices of Part B

To complete this second part of the book, we have concocted two appendices:

- **Examples of applications – to give you some application ideas.** This first appendix presents a very specific example of the approach to defining, choosing and justifying FlexRay parameters.
- **Scheduling problems – application of the FlexRay protocol to the static (ST) and dynamic (DYN) segments.** The purpose of the second appendix is to widen the technical and mathematical field of vision associated with the problems of access to the medium, and of scheduling messages transmitted during the DYN segment via the minislots.

Appendix B1

Examples of Applications

In this appendix, we have chosen to present to you a FlexRay application example[1] which is built with the aid of, and around, the one which was the first industrial implementation in a vehicle on the road, in September 2006. With no publicity intention, this vehicle was the BMW X5 (see Figure B1.1).

The BMW X5 (Development Code L6)

The X5 model SAV (sports activity vehicle) was the first vehicle partly using a FlexRay communication network to be manufactured and marketed.

In a first phase of starting up production, BMW used FlexRay in a system called 'Adaptive Drive', which acts simultaneously as an active anti-roll stabiliser and an electronic shock absorber.

It should be noted that the expression 'X-by-Wire' applies when mechanical or hydraulic control systems are replaced by fully electrical or electronic solutions, whether for steering (steer-by-wire), acceleration (throttle-by-wire), suspension (suspension-by-wire) or braking (brake-by-wire).

A Little Strategy

Let us begin by giving an overall view of BMW's solution. It can be summarised in a few strong ideas:

- To free ourselves of the speed limitations and non-real time nature of controller area network (CAN), and to avoid looking for improvements which would only be palliative and/or temporary solutions involving some sleight of hand, we will take a great step forward and go directly to a system which is suitable for future and new generations of

---

[1] The detailed description of this application was presented officially by Dr Anton Schell of BMW at the annual VECTOR Congress which took place in Stuttgart in March 2007.
Comment: For understandable teaching reasons in this book, while keeping all the basics of the subject, we have slightly modified the initial presentation of this application example.

FlexRay and its Applications: Real Time Multiplexed Network, First Edition. Dominique Paret.
© 2012 John Wiley & Sons, Ltd. Published 2012 by John Wiley & Sons, Ltd.

Figure B1.1 The BMW X5

applications, and requires high communication rates and high performance in real time with functional security:
- FlexRay;
- 10 Mbit/s bit rate;
- 100 ns bit duration.

- To reduce costs and development times and for time to market in relation to commercial demand, the system design will be such that we will be able and obliged to reuse the bricks which were made during this project for applications in a range of vehicles during future developments. The set of design parameters will therefore be conceived, chosen and built in such a way that it is kept identical and constant, so that all the electronic control units (ECUs) can easily be transferred to all future series of vehicles.
- For the moment, we are not making an X-by-Wire system on a large scale. First, we will familiarise ourselves with the FlexRay protocol and its subtleties, at both the physical and the protocol level, on one kind of dedicated application, and we will learn in theory

and practice. And in a second phase, we will switch the current vehicle architecture (based on CAN HS and LSFT) to a FlexRay backbone architecture, and finally, in a second-and-a-half and a third phase, we will go on to X-by-Wire.

So, the scene is set! It's a fine programme, isn't it?

Global View of the Parameters of the FlexRay System

How many times have we heard this sentence, from system designers or future users of FlexRay: *'Mr Paret, do you perhaps have, by any chance, at the bottom of a drawer, a framework, an Excel table, in short something ready-made to help me to define my FlexRay parameter set?'* Sometimes the answer is yes, sometimes it's no – because there isn't a complete recipe!

You should know that to define the parameters of a FlexRay system, every development manager must, of course, follow certain basic rules, and afterwards it is necessary to improvise as you go along, with whatever comes to hand. Let's give an example of a basic recipe:

- **Ingredients**
  - know almost by heart all the details of the FlexRay protocol, and the potential vexations of its physical layer;
  - choose the most representative colleagues, who are involved in the various aspects of the project (and there are a lot of them; not colleagues, aspects!);
  - define precisely the desired applications for the system (and that's not simple!).
- **Preparation**
  - stir it all up, first in a full meeting (and that's not simple either), then act alone to obtain a good creamy texture;
  - let all the ideas slowly settle down;
  - let it rest; the longer the rest, the better the result (the night always brings good advice).
- **Cooking time**
  - allow it to simmer for a good while over a low heat, stirring it from time to time to avoid sticking ... people to their own ideas;
  - then have the whole team come back to the meeting room (of course);
  - announce, insolently and arbitrarily, the choices that you yourself have made (without telling the meeting that it's the toned-down result of a first draft);
  - with crocodile tears, put a little chilli into the meeting to annoy some people – to find out where their functional limitations actually are;
  - put the new criticisms and comments through the strainer;
  - have handkerchiefs ready for the inconsolable;
  - make it lighter and add something to hold it together;
  - stir it again a little to make the sauce smooth;
  - finalise the last details;

And there it is, almost ready to serve, at least on paper.

After these technical-culinary proposals, let us return to the example of choosing parameters depending on what is wanted, using as an example those that were specifically defined by BMW for the X5.

Desired Functional Parameters

These parameters are linked to the desired applications. Very often, the constraints or requirements expressed by the various participants in the project conflict, and it is necessary to juggle with all of them together.

Static Segment

As we have pointed out several times, the aim of the ST segment is to take the most advantage of the offered bandwidth, with respect to cyclical messages of high priority for transporting deterministic data linked to the application. Firstly, therefore, it is useful to quantify the number of static slots required by the nodes which must communicate for deterministic application reasons and/or for the requirements of applications of real time type which make it necessary to transmit static frames, and thus belong to the ST segment.

Required Number of Static Slots

After going round the table, when all the requirements have been stated, it is found that:

- about 50 static slots are required to cover the whole application;
- taking account of reserve slots and/or those provided for future applications, it is decided to keep a total of 75 static slots in the ST segment.

But – because there is always a 'but' – in the course of the same discussion, some of the people around the table indicate that (for a variety of reasons) they need very different values for the repetition cycles of their messages. Putting these facts together gives:

- repetition cycles of static frames

for some messages (about 15, different or repeated) every	2.5 ms
for most of them (about 50)	5 ms
for some (about 10)	10 ms
for some (about 6)	20 ms
and more rarely (about 6)	40 ms

Said like that, it obviously raises a problem! Don't worry, later we will examine how to resolve the problem of different values of repetition cycles! Just be patient!

Maximum Number of Bytes Carried per Static Frame

For all participants, it is found now that the maximum number of bytes carried in static frames is 16 (8 words of 16 bits). Being prudent and thinking of the future, it is decided that the size of a static frame should be capable of carrying 24 (50% more anyway!).

Let us now go on to examine what is wanted for the DYN segment.

Dynamic Segment

As we have indicated throughout this book, use of the DYN segment and its method of access to the medium is principally intended for transmitting messages which are triggered by events and/or have no pure real time constraints or constraints on flexibility of operation, to avoid wasting the bandwidth of the ST segment. This can include, for instance, transmission of:

- network management messages;
- diagnostic messages;
- messages for downloading to flash memory (via the so-called transport layer of the OSI model);
- calibration data messages (XCP on FlexRay);
- and any other message of event-triggered type, with repetition cycles of 10, 20, 40 ms or even greater.

Below, we will examine how to define the hierarchy of frame identifiers in this segment.

Description and Justification of the Implemented Choice

The next few lines suggest a possible answer to the problem stated above.

General Composition of the FlexRay Communication Cycle

The total duration of the proposed communication cycle is 5 ms (we can already hear your cries about the 2.5 ms repetition cycles of some frames! Patience, my friends!), which breaks down as follows:

- an ST segment of duration = 3 ms;
- a DYN segment of duration = 2 ms – network idle time (NIT);
- no symbol window = 0 ms (so the system has no bus guardian);
- value of NIT = 100 µs.

Details of the Parameters of the FlexRay System under Consideration

Let us now go on to examine the ST and DYN segments in detail.

Static Segment

The chosen parameters for the ST segment are as follows:

- ST segment duration = 3 ms;
- number of static slots = 91.

No, this isn't a mistake, although we only asked for 75 initially! What is the hidden reason for the 91 – 75 = 16 extra static slots?

If you remember (see above), when we went round the table it was said that about 15 static messages had to have a 2.5 ms repetition cycle, and this is how the problem has been solved with a FlexRay communication cycle of 5 ms!

When we chose the total duration of a FlexRay cycle to be 5 ms and the duration of the ST segment to be 3 ms, without telling you we very hypocritically chose to break the duration of the ST segment into three parts (see Figure B1.2):

- from 0 to 0.5 ms: transmission of the family of messages which must be transmitted every 2.5 ms;
- from 0.5 to 2.5 ms: transmission of the family of messages which must be transmitted every 5 ms (or every 10, 20, 40 ms – see below);
- from 2.5 to 3 ms: transmission of the family of messages which must be transmitted every 2.5 ms;
- then the DYN segment, from 3 to 5 ms.

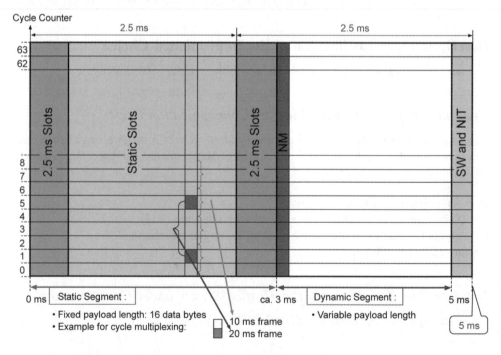

Figure B1.2 Duration of the static slot

Consequently, as the figure clearly shows, for each FlexRay communication cycle of duration 5 ms, two zones/periods of 500 µs of the duration of the ST segment (at the start and end of it), which are solely reserved for real time messages, are scrupulously repeated every 2.5 ms (obviously hoping that the CPUs can process them in this rhythm, or that this is considered and processed as information redundancy for security during a single cycle, and so on).

Let us now translate the duration of the ST segment (3 ms) into its equivalent number of bits:

$$\text{duration of a static slot} = 3000/91 = 32\,967\,\text{ns} = 32.967\,\mu\text{s}$$

This is equivalent to 329.67 bits of duration 100 ns.

The number of static slots which repeat every 2.5 ms thus equals:

$$500\,\mu\text{s}/33\,\mu\text{s} = 15\,\text{static slots}$$

which is what was wanted, leaving

$$91 - 2 \times (15) \approx 60 \text{ slots for all the other static slots}$$
$$\text{(those which repeat at 5, 10, 20 or 40 ms)}$$

Whence the famous initially requested total of $60 + 15 = 75$!

The 60 messages of which the repetition cycles are set to 5, 10, 20 or 40 ms will be distributed depending on the application wishes issued initially, using time multiplexing of the static frames in the static segments of the FlexRay communication cycles from 0 to 63 (see an example of multiplexed time distribution in Figure B1.3).

❑ **Characteristics:**

    ❑ In different cycles different frames are sent within the same slot and on the same channel

    ❑ Repetition = Repetition rate

    ❑ Base Cycle = StartCycle, Offset

❑ **Info on Frames:**

    ❑ Slot ID
       [Slot number within a cycle]

    ❑ Base Cycle [Cycle number of the first occurrence within the schedule, 0 to 63]

    ❑ Repetition [send interval in multiples of the cycle time]

    ❑ Channel

    ❑ TX nodes

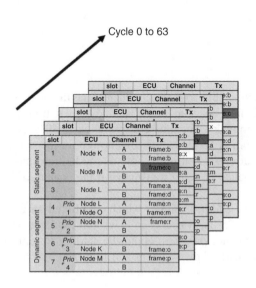

Figure B1.3 Multiplexed time distribution

Let us now translate the number of bits per static slot into the number of bytes transported per static frame. Knowing that a static frame consists of:

Header	40 bits
TSS (6–15)	11; value chosen depending on maximum required topology
BSS + FES	3
CRC	24
CID	11; => a total of 89 bits of system encapsulation

TSS = transmission start sequence; CID = channel idle delimiter.

Leaving:

=> 329 – 89 = 240 bits = 24 bytes of 10 possible bits
that is, maximum 24 × 8 useful bits = 192 bits per frame.

This meets perfectly the stated wishes for 16 bytes per static frame, with a reserve that can extend to 24.

That settles the ST segment. Let us now go on to the DYN segment.

Dynamic Segment

The chosen parameters for the DYN segment are as follows:

- DYN segment duration = 2 ms – NIT = 2 ms – 100 μs = 1.9 ms;
- initial duration of minislot = 6.875 μs = 6875 ns, equivalent to 68 bits;
- maximum number of minislots = 2000/6.875 = 290.9 = 290.

Some Comments

Let's start with some comments.

- There's no point in dreaming about being able to use 290 identifiers in the DYN segment! Just to clarify things, you should now be aware that it is impossible to transport a frame, even an empty one, in the last minislot, since its encapsulation (=40 + 24 + 11 + ... bits) already exceeds the maximum of 68 bits which are allowed for the size of this last minislot! If necessary, the last but one – if it is the only one to be transmitted during the DYN segment – will have a chance of being transmitted if the content of its frame is not too long!
- It is therefore true that being able to use the 290 possibilities which are offered because a low initial duration of the minislot, of 6.875 μs, was chosen is wishful thinking. The only importance of this value is having greater precision about the starting instant of frames, particularly when they are at the start of the segment and it is therefore more certain that they will be transmitted.

- Knowing that the longest possible FlexRay frame makes a total of 2638 bits – that is $263\,800\,\text{ns} = 263.8\,\mu\text{s}$ – during the 2 ms DYN segment, you can only hope to transport – using the first minislots – $2000/263.8 = 7$ of the longest frames, of 2032 useful bits.

The comments above lead to a further comment:

- If it is necessary to transport long frames, in a way which is not specifically repetitive or event-driven, and to be certain that they are transmitted correctly every time, it is better to put them at the start of the DYN segment. In particular, this is the case for 'network management' and/or 'flash downloading' phases, which are used, for example to reconfigure the network structure from time to time. You should therefore not be surprised to find these functions assigned to the first minislots of the DYN segment.

To conclude this quick reflection about this application example, let us examine another way of stating the above section, to be read attentively because it opens up many applications: *by arranging certain well-chosen event-driven functions in the first dynamic minislots, that is at the boundary of the ST segment, and by giving them very high priority relative to the other minislots for access to the network (despite their event-driven aspect), these minislots become practically 'static slots', with the possibility of being able to distribute one or more very large frames if necessary. These few dynamic minislots are thus seen as 'static slots' which are in the DYN segment but occur from time to time (the event-driven characteristic), but have the highest priority of the minislots.*

To complete this chapter of examples, let us give some numbers for the values of local and network parameters:

- **Local parameters** (at the level of a network node)
  – clock frequency, specific to the FlexRay part $= 80$ MHz;
  – clock period of FlexRay microcontroller $= 12.5$ ns;
  – microtick duration $= 25$ ns;
  – duration of FlexRay bit $= 100$ ns;
  – number of microticks per bit $= 100/25 = 4$.
- **Network parameters** (at the global level of a cluster)
  – macrotick duration $= 1.375$ ns;
  – number of macroticks per cycle $= 5000/1375 = 3636$;
  – number of microticks per macrotick $= 1375/25 = 55$.
- **Communication cycle timing:** – see Figure B1.2.

Appendix B2

Scheduling Problems – Application of the FlexRay Protocol to Static and Dynamic Segments

To finish on the subject of the theory of the probabilistic character and optimal assignment of the dynamic frame identifiers which govern the transmission of messages using dynamic minislots, you should know that this concerns certain kinds of mathematics, which are suitable to real time systems and their specific environments, and this is largely outside the scope of this book. To avoid leaving you completely hungry, we have decided to offer you a technical and mathematical appendix which enables you to take your first step into this field.

Introduction

As an introduction to this second technical appendix, here are a few words in the form of a WARNING.

As you will certainly have noticed, in the course of Chapter 7, concerning access to the medium, the dynamic (DYN) segment holds many mysteries concerning the probabilistic aspects and their consequences in FlexRay applications. This is why we have decided to include a specific, technically very detailed appendix, the purpose of which is to recall some truths associated with the problems and application performance of the ST segment, and more particularly of the DYN segment. In fact, this technical appendix forms a real chapter in itself. If you want to have only an overall view of the operation of the FlexRay protocol, it can be skipped, but – and it's an important but – be very aware that when you are confronted with the actual implementation of FlexRay, sooner or later, you will come back to it, and it will become simultaneously your nightmare and your bedtime reading, because the application basis of FlexRay is based on problems which are directly associated with the problems of scheduling static (ST) and DYN segments and their application consequences. You have been warned!

FlexRay and its Applications: Real Time Multiplexed Network, First Edition. Dominique Paret.
© 2012 John Wiley & Sons, Ltd. Published 2012 by John Wiley & Sons, Ltd.

One point concerning reading this appendix: don't be surprised by its form! It has intentionally been structured and presented in the form of a 'scientific publication' since the field of scheduling – for FlexRay in particular – is expanding rapidly, and to get precise ideas it is necessary to hunt for information every day, and to make summaries and validations. This is why this appendix was edited with Ms Rim Guermazi.[1]

Before starting on the basics of this subject, with which readers may be more or less unfamiliar, we offer a reminder – for once, not a brief one – to re-examine closely the concepts of real time systems, scheduling and the related terminology.

Problems of 'Real Time' Systems

Concepts of Reactive Systems and Time Constraints

In general, IT systems are divided into three classes:

- **Transformational systems** – they transform data to produce a result. These results depend only on the input data. Scientific calculation programs are an example of transformational systems.
- **Interactive systems** – they interact with their environment. These systems associate processing with the events that they receive. In most cases, the principal actor in their environment is the human user. Office automation programs are an example of interactive systems.
- **Reactive systems** – they react to stimuli from their environment. Unlike interactive systems, the environment of reactive systems evolves independently of them, and does not wait for the end of processing which was caused by the stimulus it has issued. Command and control systems are examples of reactive systems. Reactive (real time) systems are integrated with the environment that they control, and are subject to the same physical constraints (temperature, pressure, magnetic field, and so on) as their physical surroundings.

Real Time Systems and Their Classification

Many on-board systems are called real time systems. Their purpose is to monitor and drive dynamic processes. There are numerous definitions, but we propose the following: *'A real time system is one of which the correctness depends not only on the logical results of the calculations which the tasks carry out, but also on the instant when these results are produced'*. Thus, these systems are constrained to carry out a quantity of tasks within a limited time interval. Speed of execution of the software alone does not guarantee the

---

[1] I have had the opportunity and pleasure of teaching in numerous engineering schools for years. I discovered Rim Guermazi during her last year of studying the on-board systems option at ESAIP near Angers, and simultaneously finishing a master's degree in the Science faculty. Together with, on the one hand, Mr Laurent George, professor and responsible for research at ECE and well known for his publications in this field, and on the other hand Mr H. Belda of Vector, we suggested to her a doctoral thesis about the scheduling problems of FlexRay. This appendix, which was written in collaboration with her, summarises the principal existing works in this field, on a given date and at the level of this book. Thanks again Rim!

validity of the system. In other words, this definition implies that, in all cases, all the time constraints must be complied with, and otherwise the system is faulty.

Classification of real time systems is based on the possible consequences of a violation of time constraints.

Hard Real Time Systems

When a system is subject to strict time constraints (that is the time constraints must be complied with at any cost), they are called *strict* or *hard real time systems*. Violation of these constraints can have catastrophic consequences for the controlled environment. Examples of hard real time systems are found in the nuclear industry, for instance. These systems can also be classified as safety-critical systems, where a failure causes human losses, and mission-critical systems, where a failure causes economic and environmental losses.

Critical Real Time Systems

These must be validated in all operational scenarios, even in the worst cases, since a delayed result is a wrong result and the consequences of a failure exceed the added value of the system. These systems are subject to reliability constraints. This characteristic brings problems of operational safety and fault tolerance. For information, an example of a method of evaluating the probability of occurrences of failure in an automotive on-board system, and how to comply with the requirements in terms of operational safety taking account of this evaluation, was proposed a few years ago. This method consists of evaluating the worst delay which the control system can tolerate regarding the arrival of an instruction, taking account of the dynamics of the vehicle and the operational architecture. However, this method is restricted, on the one hand to the study of steering systems, and on the other hand to electromagnetic interference as the source of transient faults (faults with a limited duration and with causes depending on the environment) which can alter the system.

To support critical applications, a critical real time system must be:

- **Deterministic** – two aspects of determinism are distinguished:
  - value determinism, which implies that the same sequence of inputs will always produce the same sequence of outputs;
  - time determinism, which implies that the timing characteristics of the outputs are always preserved.
- **Predictable** – to guarantee an adequate performance level, the system must be capable of predicting the consequences of any scheduling decision which is taken. If compliance with a time constraint of a task cannot be guaranteed, the system must consider this risk in advance, to plan alternative actions in response to this event.
- **Fault-tolerant** – a hardware or software fault must not cause the whole system to fail. The hardware and software components that are used in these systems must therefore tolerate these faults.
- **Tolerant to heavy workloads** – in conditions of operation under load, real time systems must be robust. They must therefore be capable of anticipating all operational scenarios.

- **Maintainable** – the architecture of a real time system must be modular, to make it possible to make modifications easily.

Soft Real Time Systems

Another type of system, called a soft real time system, includes devices where there are lower requirements concerning compliance with time constraints. These systems can tolerate a certain threshold of time faults (that is the time constraints must be complied with as far as possible) which result in degradation of performance (quality of service). For these systems, metrics are defined to quantify the quality of the proposed services. Depending on the complexity of the system, they can be obtained by analysis (usually statistical) or by evaluation (usually by simulation). Among the most common metrics, the rate of compliance with constraints, the use threshold and system costs can be mentioned.

Complexity of Distributed Real Time Systems

The technological progress which has been achieved in the electronics field (robustness, stability, immunity to variations of electromagnetic fields, and so on) and the constant reduction of the price of components have contributed greatly to the development of on-board real time systems in industry. This fact has contributed greatly to the development of systems, in the automotive field in particular, where functions which are impossible to obtain in a mechanical system have been implemented: assisted braking, electronic braking distribution, assisted steering, and so on. We will now concern ourselves with the advantages of distributed architectures and the issues about using them.

Characteristics and Advantages of Distributed Systems

As a reminder, a distributed system consists of hardware components (several computers which share neither physical memory nor a physical clock) and software components (algorithms which are executed on the various computers), which communicate and coordinate their actions via messages. The complexity of the processes to be monitored and/or controlled, the quantity of data and events to be processed, the geographical distribution of the hardware elements of the system to be controlled and the appearance, several years ago, of communication protocols have contributed to the importance of distributed systems. This architecture makes it possible to overcome the limitations of centralised systems. In fact, distributing the execution of tasks makes possible faster (parallel) processing than in single-processor systems (sequential processing). Additionally, in safety systems where it is necessary to guarantee the tolerance of the system to faults, distributed architectures make it possible to use redundancy in both hardware and software.

Heterogeneity in Distributed Real Time Systems

The principal reason for the complexity of management of distributed real time systems is the heterogeneous nature of its components:

- different computers and microprocessors with different technologies;
- hard and soft time constraints which can coexist in the same system;
- coexistence of analogue and digital components, and so on.

Additionally, the interaction of a real time system with its environment can be classified according to the 'event-triggered' and/or 'time-triggered' paradigm:

- In a time-triggered system, all activities are executed at predetermined instants. It is said that the cadence of these systems is set by time.
- In contrast, an event-triggered system implements the notification of new events by an interrupt mechanism, where the system reacts to these stimuli as quickly as possible. The cadence of these systems is set by their environment. In fact, changes in the environment raise interrupts, and call up the mechanisms for managing them. These systems have the advantage of reacting immediately to stimuli, but it is more difficult to guarantee and prove that they function well in overload scenarios. Additionally, in these systems the execution medium must carry out actions such as recording the current context and restoring it afterwards, whereas time-triggered systems require only the maintenance of a global clock for synchronisation. Time-triggered systems themselves introduce latencies, but their operation is predictable. They are therefore implemented in processes where reliability of operation must be ensured (critical systems concerning the safety of human life, for instance). Event-triggered systems are suitable for command and control systems.

In this appendix, which is dedicated to applications of the FlexRay protocol, we will consider systems where time-triggered and event-triggered actions coexist (see Figure B2.1).

In a distributed system, the various nodes communicate via a communication network, using messages. The network thus forms a resource which is common to all the nodes, as in single-processor systems where tasks share the processor time. In distributed systems, it is the medium access control (MAC) layer of the OSI model which is responsible

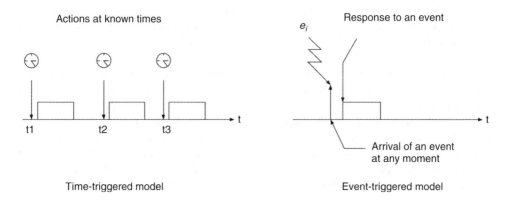

Figure B2.1 Event-triggered model versus time-triggered model

for how a message accesses the network. Next, we are more particularly interested in the FlexRay communication protocol, which was developed for the automotive sector. In what follows, therefore, we will present the particular features of this protocol in relation to its policy for access to the medium.

FlexRay

As was explained at length above, FlexRay defines a deterministic, multi-master network which makes it possible to combine synchronous and asynchronous transmission by combining transmission of time-triggered and event-triggered messages. This protocol makes it possible to support numerous topologies, and makes redundant transmission possible thanks to two transmission channels. In Chapter 14, we will show that synchronisation is based on a distributed mechanism for synchronising the local clocks, where each node synchronises itself relative to the others using a fault-tolerant synchronisation algorithm.

The FlexRay communication method is based on periodic repetition of a communication cycle (*gdCycle*). This cycle consists of an ST segment, followed by a DYN segment, and is terminated (optionally) by a symbol window and a network idle time (NIT). The NIT field enables the various nodes of the network to synchronise with each other. In the rest of this appendix, we will disregard the last two fields, because they are not included in the technical scope of scheduling problems. The two segments (ST and DYN) which are defined in a FlexRay communication cycle consist of a number of slots:

- static slots, characterised by *gNumberOfStaticSlots*, which fixes the number of slots in the ST segment;
- the parameter *gdStaticSlot*, which fixes the size of a slot (all the static slots are the same size);
- the DYN segment consists of a number (*gNumberOfMinislots*) of minislots, each of a duration *gdMinislot*.

The slot counter in the DYN segment may not reach its maximum value, thus forcing low priority messages to wait for the next communication cycles to try to access the bus.

The aim of the network designer is therefore to study and propose a method of assigning priorities (frame ID) to dynamic messages, to guarantee compliance with their time constraints. In what follows, we will consider only deadline constraints in our explanations.

In a distributed system, the communication medium (bus) can be assimilated to a processor where the tasks (messages) try to access it. We will therefore be inspired by scheduling algorithms which exist and are used for single-processor systems, to identify the most appropriate type of scheduler for our problem. In this way we will approach the problems of scheduling real time systems, and more particularly we will spend some time on scheduling single-processor systems and scheduling communication in distributed systems. To begin with, we will now be particularly interested in the policy for access to the medium in this network; that is, the MAC layer of the OSI model.

Scheduling Real Time Systems

In real time distributed systems, many tasks try to access shared resources such as processors and networks simultaneously. Scheduling tries to solve the problem of efficient use of these resources to guarantee that the system is correct, in other words that it complies well with all its time constraints.

Compliance with time constraints in a real time system is formally verified thanks to sizing. The execution support is responsible for distributing the requests for resources, while ensuring compliance with all the constraints. It can take the form of an operating system which controls the hardware directly, or of middleware which is inserted between the operating system and the application.

The special feature of real time execution support is that it takes account of the time constraints and of the behaviour over time of the application and system. In particular, it is a component of the execution support that decides which task must be executed on which processor: **the** *scheduler*. **It will be the cornerstone of communication management in FlexRay!**

The following paragraphs present a summary and analysis of scheduling techniques for real time systems, and of different approaches which exist in the state of the art. We will therefore now:

- present the problems of scheduling, and the various concepts that are attached to it;
- analyse the most common approaches for single-processor systems; particular attention will be given to non-preemptive, fixed-priority (FP) algorithms;
- in this context, detail how the worst-case scenario and the associated feasibility conditions are obtained; this choice is justified by the fact that communication in the DYN segment of the FlexRay protocol is based on fixed priorities (frame ID), and that a message which is accessing the bus cannot be preempted even by a higher priority message;
- be interested in the problems of scheduling in distributed systems, where the different approaches to scheduling communication of messages will be presented.

Scheduling and Analysis of Schedulability

A scheduling algorithm supplies the sequence in which the tasks access the various resources of the system such as processors and networks.

Scheduling for a set of tasks is said to be *feasible* if execution of all these tasks complies with their constraints such as their deadlines and precedence relationships. A set of tasks is said to be *schedulable*, or *feasible* again, if at least one algorithm which produces feasible scheduling exists. *Optimality* in scheduling theory has a different meaning from what is found in the field of applied mathematics. In fact, a policy P is said to be *optimal* in relation to a class of scheduling problem if, given that scheduling of a set of tasks is feasible with a given policy, it is necessarily also feasible with P. Consequently, if a system cannot be scheduled by an *optimal scheduler* of a given class, it cannot be scheduled by

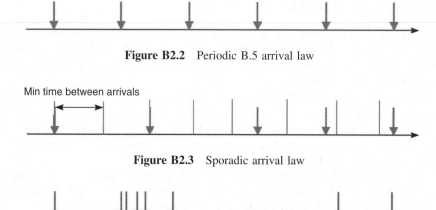

Figure B2.2   Periodic B.5 arrival law

Min time between arrivals

Figure B2.3   Sporadic arrival law

Figure B2.4   Aperiodic arrival law

any algorithm of the same class. However, the problem consisting of constructing feasible scheduling can be reduced to a 'classic' optimisation problem. In fact, certain propositions have concerned the minimisation of a criterion such as the maximum delay of tasks. Thus, once the maximum delay has been minimised, proving that the system complies with all its time constraints comes back to verifying that this maximum delay is negative or zero (Grenier, 2004).

The method which is generally used to solve a scheduling problem begins by identifying **the class of real time problem**: characterising the *task model*, the *time constraint model* and the *scheduling model* that are used.

That is what we are going to do now – and it's what you will have to do when you develop your project, to define properly the type of scheduler to be used according to your application.

Class of a Scheduling Problem

The Task Model

The execution support of real time systems must know the timing characteristics of the tasks to guarantee compliance with their time constraints. These characteristics form part of the *task model*, which usually involves:

- **The task arrival law** – this is about how requests for activation of a task τ_i are repeated over time. We now distinguish:
  - **Periodic arrival**: the requests for activation of a task are periodic, with period T_i.
  - **Sporadic arrival**: the requests for activation of the task are such that a minimum duration separates two successive arrivals (minimum inter-arrival $\geq T_i$).
  - **Aperiodic arrival**: as the privative 'a' indicates, the arrival of a task cannot be characterised by any law, and it can arrive at any instant. A sporadic task is a special case of an aperiodic task.

- **Arrival on a sliding window**: where there are at most n_i arrivals of a task on a sliding time window W_i Laurent George (2005) states that this model should be studied using the sporadic model, specifying that n_i arrivals on a window W_i are equivalent to n_i independent sporadic tasks of period W_i.
- **The activation instants of a task** – when a particular activation scenario is imposed on the tasks, the tasks are said to be *concrete*. Similarly, if no assumptions are made about these activation instants, the tasks are said to be *non-concrete*. For the periodic arrival law, for instance, considering a concrete task model means defining the first activation instants of the tasks.
- **Execution times** – this is the execution time of the task itself in the processor. In general, the worst-case execution time (WCET or C_i), which forms an important input parameter for the methods and tools for verifying compliance with the time constraints associated with tasks, is quantified.
- This WCET is an upper bound on all the possible execution times of a task by itself, with no preemption by the operating system. This parameter is determined either by analysis of the machine code of the task and of the hardware architecture on which it is executed, or by a benchmark on the real architecture.
- **The response time** – this is the time which separates the request for activation of a task from its end of execution. Here, too, we try to characterise the worst-case response time (WCRT or r_i), which represents the longest interval of time separating the request for activation of the task from its end of execution. This parameter depends on the scheduling algorithm which is used, and takes account of the delay introduced by tasks of higher priority than τ_i. It is always true that $r_i \geq C_i$.

Time Constraints

A real time application consists of tasks which are executed in parallel. These tasks can interact with each other via synchronisation or mutual exclusion mechanisms or precedence relationships. In this case, they are called 'interdependent' tasks.

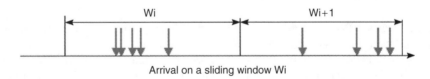

Arrival on a sliding window Wi

Figure B2.5   Arrival on a sliding window W_i

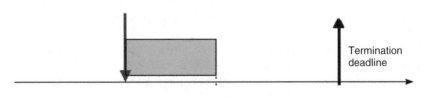

Termination deadline

Figure B2.6

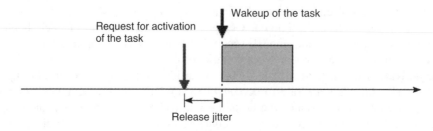

Figure B2.7

Scheduling is simpler in systems where the tasks are independent, since in this case the order in which they are executed is unimportant. The task model also makes it possible to express time constraints relative to the tasks. The commonest time constraints are:

- **The termination deadline** – this is the instant when the task must be terminated. This deadline can be absolute or relative to the creation time of the task. The constraint on the deadline of a task can also concern its start time.
- **Jitter** – this is the time interval which separates a request (an event) and the instant when it is acted on by the system. It is called 'release jitter' when the event concerns a request for activation of a task.
- **Precedence** – the case in which starting a task is constrained to wait for the result of another task is called the precedence constraint. This constraint is represented using oriented graphs, in which the vertices represent the tasks and the arcs represent the precedence relationships.

Models of Schedulers and Scheduling

A scheduler is said to be 'non-idle' if the processor is never inactive when tasks are ready to be executed. In this case, it is said to function without empty time. Otherwise, it is said to be 'idle'.

Schedulers can also be classified as follows:

- **Online versus offline**: On the one hand, in offline algorithms, all decisions are made at compilation time, and are stored in a table (static plan). During the execution of the system, no scheduler is required. These algorithms require a priori knowledge of the timing properties concerning all the tasks. On the other hand, in online algorithms, all decisions are made during execution of the system, each time a task arrives or another completes its execution. The properties of the tasks are known and taken into account at this instant. These schedulers can be based on the construction of a scheduling table in the course of operation (dynamic plan), or more commonly they are based on the concept of priority.
- **Preemptive versus non-preemptive**: A 'preemptive scheduler' can interrupt a task in the course of execution, when another, of higher priority, arrives. On the other hand, a 'non-preemptive scheduler' executes the task until it terminates, without concerning itself with the possible demands of tasks with higher priority. Some tasks explicitly require non-preemptive execution. For instance, we can mention the interrupt manager,

which is responsible for recording the state of the system each time an interrupt is going to be processed. If the application does not explicitly require restrictions on preemption, it is not easy, in the general case, to decide on the better alternative.

- **Centralised versus distributed**: In a distributed system, the scheduling is said to be 'distributed' if the scheduling decisions are made by a local algorithm at each node of the system. It is called 'centralised' if the scheduling algorithm is executed on a single, privileged node for the whole system.
- **Static versus dynamic**: In 'static' scheduling, all scheduling decisions are based on fixed parameters which are assigned to the tasks before they are activated, whereas in 'dynamic' scheduling, all decisions are made in relation to dynamic parameters, which can change while the system is running. Static scheduling is less flexible than dynamic scheduling, since it requires a priori knowledge of all the attributes of the tasks. However, although dynamic scheduling makes it possible to take account of unpredicted events, and to use the processor better, it requires more resources, in particular concerning memory requirements, than static scheduling.

Static and dynamic scheduling have yet another meaning in multiprocessor architectures and distributed systems. In a static system, the tasks are grouped into subsystems and allocated statically to the processors. Tasks at the level of a processor are then scheduled independently at that level, except in scenarios where distributed tasks must be synchronised. In a dynamic system, the tasks are allocated dynamically to the available processors in the system. In this case, a common queue for all processors is created, and the task at the head of the queue is allocated to the first processor which is in idle mode. If preemptions are permitted, a task can migrate from one processor to another to complete its execution.

Online and offline scheduling models can both use a preemptive or non-preemptive scheduling algorithm. Offline scheduling algorithms imply static scheduling, in which the dynamic activities (or tasks) are incorporated by a change in the scheduling table or by use of a hybrid scheduler (a combination of the two models, online and offline).

Online scheduling algorithms can be static or dynamic from the point of view of assigning priorities to tasks. Additionally, certain hard real time applications require tasks to be created and destroyed dynamically, according to the requests from the environment. These requests cannot be predicted and taken into account in an offline analysis of schedulability. These dynamic systems require an online scheduling algorithm, and tests of schedulability which can be applied online. An example of these applications is air traffic control, where each aircraft is controlled by a task, and the number of these tasks depends on the number of aircraft which are passing through the controlled zone.

Different Approaches to Real Time Scheduling

The approaches which are most used for scheduling real time systems can be classified as follows:

- clock-driven approaches;
- round robin (RR);
- scheduling based on priorities.

They differ in how the scheduler is implemented, and in when its decisions are made. These approaches can also be used to assign tasks to the processor(s) as well as to schedule communications in distributed architectures. The term 'task' will therefore be used indifferently to designate a system task or a communication frame, and the term 'processor' will be used to designate an active resource in the system.

Clock-Driven Scheduling

The scheduler's decisions are made at precise instants, independently of events such as requests to activate new tasks. The term 'time-driven' scheduler is also used. The parameters of the tasks are fixed and known, and the scheduling is done offline. A disadvantage of this approach is the necessity of knowing, a priori, all the instants of requests for activation of all the tasks, to be able to schedule them (the case of concrete tasks). This algorithm is less flexible than those that are based on dynamic properties, but it ensures that the system is predictable. This type of scheduler is most often implemented as follows: decisions are made periodically, and a timer is included so that the scheduler can be called up.

Round Robin Scheduling

The RR scheduler is of preemptive type, and is based on an approach of sharing the processor time. It executes each task during a quantum which is associated with it, and if it has not completed its execution, it is placed at the end of a first in, first out (FIFO) queue and waits its turn to continue execution. The algorithm is defined by the Posix 1003.1.b standard. This approach is typically used for scheduling communication of messages, but it is not appropriate to scheduling real time systems because of the execution delays to which each preempted task must be subjected, consequently increasing its response time. In practice, this type of scheduler is used for low-priority tasks which require a long execution time, such as diagnostic tasks. A variant of this method, weighted fair queuing (WFQ), has been proposed, and assigns different time quanta to tasks, so that the tasks do not have the same share of the processor time.

Scheduling Based on Priorities

Scheduling based on priorities is the most classic, the most studied and the most used technique in the field of real time provisioning, and there are numerous results in the literature. The scheduler assigns priorities to the instances of tasks, on the basis of static or dynamic properties. It thus maintains an ordered queue of ready tasks. Each time it has to make a decision, it schedules the task with the highest priority (HPF, highest priority first). Decisions are made when a task completes its execution and another receives a request for activation. The scheduler is of online, event-driven and non-idling type.

In this category, algorithms based on assignment of fixed priorities are distinguished from algorithms based on assignment of dynamic priorities. These are the most studied approaches, more particularly in the case of single-processor systems. In an FP scheduling algorithm, all the instances of a task keep the same priority for as long as the system is in operation, whereas in a dynamic priority algorithm, the scheduler considers dynamic parameters, and consequently assigns different priorities to the various instances of a task. Algorithms based on priorities can also be classified as preemptive and non-preemptive.

In practice, analysis of the schedulability of a real time system is based on one of the three following approaches:

- the approach based on mean analysis;
- the approach by simulation;
- the worst-case approach, which differs from the others because it makes it possible to guarantee that for any possible operational scenario, the tasks will comply with their time constraints.

Before we continue, let's introduce some definitions concerning the processor activity.

Use Factor of the Processor

This factor represents the percentage of use of the processor to execute a set of tasks $\tau = \tau_1 \ldots \tau_n$ (Leung and Whitehead, 1982). This factor is defined by:

$$U = \sum_{i=1}^{n} \frac{C_i}{T_i}$$

An obvious necessary condition is $U \leq 1$.

Processor Demand

This parameter, $h(t)$, represents the sum of the execution times required by all the tasks in the synchronous scenario where the instant of activation and the absolute deadline are in the interval $[0, t]$. It is a measure of the minimum quantity of work which must be executed to comply always with the task deadlines:

$$h(t) = \sum_{D_i \leq t_i} \left(1 + \left\lfloor \frac{t - D_i}{T_i} \right\rfloor \right) \times C_i$$

Workload Requested by a Synchronous Task T_i

In the interval $[0, t]$, the workload is given by the expression:

$$\left\lceil \frac{t}{T_i} \right\rceil \times C_i \text{ and in the interval } [0, t] \text{ is } \left(1 + \left\lceil \frac{t}{T_i} \right\rceil \right) \times C_i$$

We will now examine various scheduling algorithms which exist and are used for single-processor distributed systems of FlexRay type. However, the aim of studying scheduling algorithms in single-processor systems is to be inspired by the results of these policies to solve the problems of allocating priorities for dynamic messages in the FlexRay communication protocol.

Scheduling in Single-Processor Systems

In this section, we will concern ourselves with clock-driven algorithms and algorithms with priorities for scheduling tasks in single-processor systems:

- In the case of clock-driven scheduling, the scheduling table is created offline, and an analysis of schedulability is unnecessary in this case, since the scheduling table is constructed in such a way that all the constraints of the system are complied with.
- In contrast, algorithms based on priorities require an offline analysis of the schedulability (trying to make the system feasible), to guarantee that the system will comply with all its constraints when it is designed.

Clock-Driven Scheduling

In the case of periodic concrete tasks, it is possible to construct a static offline scheduler which specifies the instant of execution of each task in the system. In this case, the scheduler allocates to each task τ_i a processor time equal to its corresponding C_i (WCET). In this case, the guarantee of compliance to the time constraints of the system is conditional on the creation of such a scheduler.

Most clock-driven scheduling algorithms consider a set of synchronous tasks, and the scheduler is constructed on the hyperperiod of the set of tasks (lowest common multiple of all the periods T_i of the set of tasks $\tau = \tau_1 \ldots \tau_n$), and is periodic: cyclic scheduler (Lawler and Martel, 1981).

The scheduler is most often a table where each entry specifies the instant when the scheduler makes a decision and the corresponding task which must be executed. An interrupt of Timer type will call up the scheduler at the proper time.

If the tasks are independent, the scheduler can be based on one of the algorithms with priorities such as earliest deadline first (EDF). On the other hand, in the case of a set of interdependent tasks, the scheduling problem is NP-complete. Additionally, offline schedulers can use preemptive algorithms or, in general, any algorithm which would reduce the load on the processor. Thus, in this case the scheduling is most often obtained thanks to the 'branch and bound' method, to find feasible scheduling, and heuristics are used to minimise the search space.

Algorithms Based on Priorities

Algorithms with Fixed Priorities

'Rate Monotonic (RM)' Scheduling

The priority of a task is a function of its period T_i; the shorter the period, the higher the priority of the task. This algorithm is not optimal in general; in particular, it is not optimal in a non-preemptive context or if the periods and deadlines are independent. It is optimal if $T_i = D_i$ for preemptive sporadic or periodic tasks. Since 1973, the upper limit of the use factor of the processor for the RM algorithm has been estimated as follows (Leung and Whitehead, 1982):

For n tasks: $U = n \left(2^{1/n} - 1 \right)$, and therefore, for an infinite number of tasks,

$$U \underset{n \mapsto \infty}{\longmapsto} \ln 2$$

This limit is applicable in single-processor systems.

Originally, this algorithm was applied to independent periodic tasks. Since then, RM has been generalised for analysing the schedulability of aperiodic tasks and for analysing the scheduling of communications, in particular in a token ring network, so that it is now called generalised rate monotonic (GRM). Later, on the one hand the possibility of applying RM in a distributed context, and on the other hand the consequences and additional constraints introduced by synchronisation of tasks were studied, and it was shown that when $D_i \leq T_i$, RM can be applied by adding the access blocking time of the tasks to the resources (introduced by the mutual exclusion mechanisms) to the execution times of these tasks.

'Deadline Monotonic (DM)' Scheduling

The DM algorithm (Leung and Whitehead, 1982) was proposed as an extension of the RM algorithm. In the DM policy, the priority of a task is a function of its relative deadline D_i; the nearer the deadline, the higher the priority of the task. This algorithm is not optimal in general. It is optimal for preemptive scheduling of sporadic or periodic tasks, if for all tasks their relative deadlines are less than their periods ($D_i \leq T_i$, $\forall\, i = 1 \ldots n$). Optimality is also guaranteed in a non-preemptive context, if $\forall\, i = 1 \ldots n$, $D_i \leq T_i$ and $\forall\, (i, j)$, $C_i \leq C_j \Rightarrow D_i \leq D_j$, but it is not general in a non-preemptive context.

'Fixed Priority with Highest Priority First (FP)'

When there is no obvious relationship, for all tasks, between their periods and their deadlines, or the priorities are imposed, this algorithm can be a solution. In fact, it is optimal for sporadic and periodic tasks in a preemptive context, and has been shown to be optimal in a non-preemptive context (George, 2005). This algorithm is based on calculating the maximum response time of a task. The algorithm examines the current set of tasks for whether a schedulable task exists. If so, it assigns the current priority to it. If no task is found, there is no solution to the problem.

Worst-Case Scenarios and Feasibility Tests in Non-Preemptive FP Scheduling

The Active Period

The concept of the active period is the basis of most feasibility conditions for single-processor real time scheduling, and it is closely linked to the concept of the idle instant.

- **Definition 1**: An idle instant is defined as being an instant t such that there are no longer any tasks which were activated before t and not completed before t.

- **Definition 2**: An active period is a time interval $[a, b[$ such that a and b are two idle instants and there is no idle instant in $]a, b[$.

The first active period of the synchronous scenario when the tasks are activated with their greatest density (periodic) is the longest possible active period. Let L be the duration of this active period. L is the solution of:

$$L = \sum_{i=1}^{n} \left\lceil \frac{L}{T_i} \right\rceil \times C_i$$

This equation can be solved by searching for the first fixed point of the series:

$$\begin{cases} L^{m+1} = \sum_{i=1}^{n} \left\lceil \frac{L^m}{T_i} \right\rceil \times C_i \\ L_0 = \sum_{i=1}^{n} C_i \end{cases}$$

The Worst-Case Scenario

In a non-preemptive context, a task which has begun execution can no longer be interrupted. We then try to calculate its starting time. If $\bar{W}_i(t_i)$ is its worst-case starting time, the response time of τ_i activated in t_i is $\bar{W}_i(t_i) + C_i - t_i$. P_i is the priority of task τ_i; remember that the highest priority has the lowest value. The calculation of the response time is based on the following properties:

- **Property 1** (Leung and Whitehead, 1982): The worst-case feasibility conditions are obtained when the tasks are at their maximum (periodic) density.
  In the case that no relationship between the period T_i and the deadline is imposed, the concept of active period of level P_i is defined (Lehoczky, 1990). This period defines the necessary duration of study to calculate the WCRT.
- **Property 2** (George, 2005): The WCRT of a task τ_i is obtained for FP in a non-preemptive context in the first active period of priority P_i of the scenario where all the tasks of priority greater than or equal to τ_i are synchronous at the start of the active period, and a lower priority task of maximum duration is activated one clock tick before the start of the active period of level P_i. L_i is then the solution of:

$$L_i = \sum_{\tau_j \in hp_i \cup sp_i \cup \{\tau_i\}} \left\lceil \frac{L^i}{T_i} \right\rceil \times C_i + \max^*_{\tau_k \in \overline{hp_i}} (C_k - 1)$$

where $hp_i = \{\tau_j, j \in [1, n]$ such that $P_j < P_i\}$: the set of higher priority tasks than τ_i, $\overline{hp_i} = \{\tau_j, j \in [1, n]$ such that $P_j > P_i\}$: the set of lower priority tasks than τ_i, and $sp_i = \{\tau_j, j \in [1, n], j \neq i,$ such that $P_j = P_i\}$.
For the lowest priority task τ_i, in a preemptive or non-preemptive context, $L_i = L$. To find the worst response time of τ_i, it is necessary to test, in its worst-case scenario, the activations of τ_i in $0, T_i, 2T_i, \ldots \left\lfloor \frac{L^i}{T_i} \right\rfloor$. This property is the basis of the feasibility tests presented below.

Conditions of FP Feasibility

If r_i is the worst response time of a task τ_i, a necessary and sufficient condition of feasibility is:

$$\forall i = 1 \ldots n, r_i \leq D_i \text{ and } U \leq 1$$

This necessary and sufficient condition is valid in preemptive and non-preemptive contexts.

Feasibility Test

Let us now concern ourselves with the necessary and sufficient conditions of feasibility. These feasibility conditions consist of calculating, in the first active period of priority level P_i, the successive start of execution instants $\bar{W}_i(t)$ of τ_i, activated at the instants t in $0, T_i, 2T_i, \ldots \left\lfloor \frac{L^i}{T_i} \right\rfloor$.

Theorem: the worst response time r_i of a task τ_i which is scheduled by FP is the solution of:

$$r_i = \max_{t \in s}(\bar{W}_i(t) + C_i - t)$$

where $\bar{W}_i i(t)$ is the solution of:

$$\bar{W}_i(t) = \left\lfloor \frac{t}{T_i} \right\rfloor \times C_i + \sum_{\tau_j \in hp_i} \left(1 + \left\lfloor \frac{\bar{W}_i(t)}{T_j} \right\rfloor\right) \times C_j + \max^*_{\tau_k \in \overline{hp_i}}(C_k - 1)$$

with $S = \{kT_i, k = 0 \ldots K, k \in \aleph\}$, where K is such that

$$\bar{W}_i(kT_i) + C_i \leq (K + 1)T_i$$

In other words, the task activated in KT_i terminates after its next activated request in $(K + 1)T_i$.

Given that we are only interested in FP, non-preemptive algorithms, we will not present the worst-case scenarios and feasibility conditions corresponding to the other scheduling models. The reader can consult *Conditions de faisabilité pour l'ordonnancement temps réel préemptif et non préemptif* (George, 2005) for the other cases (preemptive DP, non-preemptive DP, and so on).

Algorithms with Dynamic Priorities

The RR policy is generally considered to be a dynamic scheduling policy, even if the priority is not explicitly used by the scheduling algorithm. This choice is justified by the necessity of a priority for establishing the feasibility conditions associated with the policies. However, we will not follow this classification, and will consider the RR policy separately.

For algorithms with dynamic priorities, the priority of a task τ_i activated at instant t_i is defined at an instant $t \geq t_i$ by P_i (t, t_i).

Earliest Deadline First (EDF) Scheduling

The priority of a task is defined by its absolute deadline for termination at the latest. Thus, the priority of task τ_i, activated at instant t_i, is $P_i\,(t, t_i) = t_i + D_i$. The task with the highest priority is therefore the one with the earliest absolute deadline. This policy is optimal for preemptive sporadic or periodic tasks, or in a non-preemptive context for non-concrete tasks; it is not optimal for preemptive concrete tasks.

First In, First Out (FIFO) Scheduling

As for the RR policy, priority in FIFO is implicit. It corresponds to the instant of activation of the task. Thus, for a task τ_i activated at instant t_i, its priority is $P_i\,(t, t_i) = t_i$. This policy is optimal for scheduling non-concrete tasks when the tasks have the same deadline. In this case, FIFO behaves like EDF. However, this algorithm is not optimal for scheduling periodic or sporadic tasks.

Least Laxity First (LLF) Scheduling

The highest priority task is the one with the least laxity. Laxity is defined as the maximum waiting delay, before execution, that a task can tolerate while guaranteeing compliance with its time constraint (deadline). The priority of a task τ_i activated at instant t_i is $P_i(t, t_i) = t_i + D_i - (t + C_i(t))$, where $C_i(t)$ represents the remaining duration of execution of task τ_i at instant t_i. This policy is optimal for scheduling periodic and sporadic preemptive tasks. It is not optimal in a non-preemptive context.

All the algorithms presented above give results for periodic or sporadic tasks, but not for aperiodic (unpredictable) tasks. Since they have no constraint on their arrival, they are badly characterised. However, it is necessary to associate a priority with them, without disadvantaging the periodic tasks and without subjecting them to a famine. In the literature, four main scheduling techniques for taking account of aperiodic tasks are found:

- processing in underground tasks;
- the scanning server;
- the deferred server;
- the sporadic server.

Scheduling Communications in Distributed Systems

Scheduling in distributed systems includes, on the one hand, allocating tasks to the various processors which are available in the system and, on the other hand, scheduling communications. In a similar way, the network can be seen as a processor, and the messages as tasks, for provisioning distributed real time systems. The messages themselves can be classified as periodic, sporadic or aperiodic according to their arrival law. A deadline, representing the instant by which the message must arrive at the destination at the latest, is associated with each message. Given that the network is a common resource, a scheduling policy must specify which message has access to the network at any instant. In our study, we are more particularly interested in scheduling messages in distributed architectures. Thus, after a brief presentation of the problem of allocating tasks to processors in a distributed system, it is necessary to be interested in the different policies for scheduling communications.

Problem of Task Allocation in a Distributed System

In distributed systems, an additional constraint is added to the various constraints which have been mentioned previously: minimisation of communication times on the network. Static allocation of tasks to processors must take account not only of the costs of communication, but also of the various fault-tolerance mechanisms within the application (for example, allocation of the different replies of a message to different processors). The costs of communication must also be taken into account when the feasibility of the system is tested. Minimisation of these costs is an NP-complete problem.

The fault of much of the discussion of real time scheduling for distributed systems is the fact that the authors separate the process of allocating tasks to processors from the process of scheduling communications. This approach is used in the real time kernel MARS. Consequently, these approaches can fail to find a solution to the problem when one exists. In the next section, we will see an analysis method which makes it possible, in the case of distributed systems, to model the dependencies that exist between the system tasks and the communication messages.

Scheduling Communications

In a distributed system, the messages are subject to several delays:

- **Production delay**: time taken by the producing task to generate the message.
- **Medium access delay**: time that a message which is ready at a node spends waiting to access the medium.
- **Transmission delay**: time for transmitting the message on the network.
- **Delivery delay**: time that a message spends at the destination node waiting to be received by the consuming task, and so on.

The total delay that a message experiences is called the end-to-end communication delay.

For messages that have hard time constraints, the end-to-end delay must be included in the time analysis of the system. Two approaches can be used to determine this end-to-end delay: the stochastic approach and the deterministic approach.

- A stochastic approach studies the average behaviour of the network and is based on statistical criteria. This approach is suitable for systems with flexible time constraints, where the aim is to guarantee a rate of compliance to the deadlines.
- In contrast, a deterministic approach is based on analysis of the behaviour of the system in a worst-case scenario. This approach is particularly suitable for hard real time systems, where all the time constraints must be complied with. The worst-case durations can be obtained using either the holistic approach or the approach by trajectory. The approach by trajectory considers only the possible scenarios once the trajectory

of the communication flow is fixed. This trajectory is an ordered sequence of nodes visited by the flow. However, this method is difficult to implement in complex systems. On the other hand, the holistic approach considers all the possible worst-case scenarios. The term holistic analysis means taking account of the dependency between the scheduling of tasks at the level of the operating system and the scheduling of messages in distributed real time systems. In fact, the output time of a message is closely linked to the response time of the outputting task. Similarly, the wakeup time of a receiving task depends on the response time of the message to be received. In this context, jitter makes it possible to model the dependency of the joint scheduling of tasks and messages. Holistic analysis makes it possible to model and estimate this jitter by solving systems of recursive equations which estimate the response times of the tasks and messages.

However, calculating this duration in the worst-case scenario is not possible for all network protocols. A solution to this problem is, for instance, to preallocate the network to the different nodes of the system; the case of TDMA protocols. Below, we will discuss two approaches which are used for scheduling communications: clock-driven scheduling and scheduling with fixed priorities. However, there is a third approach called the token approach; it is outside the subject of this appendix, and we will not present it.

Clock-Driven Approach in Communications

As in the case of single-processor systems, clock-driven scheduling is based on an algorithm which is executed offline. This approach is used by TDMA protocols. The communication architecture is subdivided into time slots. Each node has a number of slots assigned to it, and it has the right to transmit on the network only in these slots. TDMA protocols require a static scheduling algorithm and setting up a mechanism to synchronise the local clocks.

Certain TDMA protocols also assign messages to slots (this is the case in the ST segment of FlexRay). Thus, the scheduling table is generated offline and guarantees the behaviour of the system throughout its development. For additional information, some TDMA protocols do not allow messages to be assigned to slots. In this case, different messages from the same node can share the same slot. In this configuration, these messages are placed in an FP queue. To refer to a method of analysing the performance of a simple TDMA protocol, see *Guaranteeing hard real time end-to-end communications deadlines* (Tindell, Burns and Wellings, (1991)). In *Holistic Schedulability Analysis for Distributed Hard Real Time Systems* (Tindell and Clark, 1994), the authors have extended the previous analysis to take account of scheduling tasks and communications, and propose a holistic analysis of scheduling a distributed real time system, based on a TDMA protocol. Their idea is based on estimating the release jitter of a task, which they define as the worst-case delay between the arrival of a task and the instant when the processor takes account of it. They estimate that the duration of this jitter depends on the necessary communication time to send a message from the source node to the destination node. Thus, if sending a message m from a node s (sender) to a node d (destination) is considered, the jitter is defined by:

$$J_{d(m)} = r_{s(m)} + \underbrace{a_m + r_{deliver}}_{C_m: \text{communication delay of m}} + T_{tick}$$

where

$r_{s(m)}$ is the response time of the process which is responsible for sending the message m,
$r_{deliver}$ is the response time of the process which delivers m to the destination processor,
T_{tick} is the jitter introduced by the Timer (the tick of the scheduler),
a_m is the worst-case arrival time of the message m at the controller of the destination node (depending on the access delay and the propagation delay of the message m).

$$a_m = \max_{q=0,1,2,\dots} (\underbrace{w_m(q) - qT_m}_{\text{access delay}} + \underbrace{X_m(q)}_{\text{propagation delay}})$$

where

q is the number of requests to send message m,
T_m is the period of message m,
$w_m(q)$ is the waiting delay of the message before accessing the communication medium.

$$w_m(q) = \left\lceil \frac{(q + 1)P_m + I_m(w_m(q))}{S_p} \right\rceil T_{TDMA}$$

where

P_m is the number of packets of message m,
S_p is the size of the slot assigned to message m (in packets),
T_{TDMA} is the duration of a TDMA cycle,
I_m is the number of packets that precede message m in the queue:

$$I_m(w) = \sum_{\forall j \in hp(m)} \left\lceil \frac{w + r_{s(j)}}{T_j} \right\rceil P_j$$

The above analysis has been extended in the case of the TTP protocol (Pop et al., 2006), and four possible policies for assigning messages to slots have been considered: static (or dynamic) assignment of a single message per slot, and static (or dynamic) assignment of multiple messages to the same slot.

For instance, let us consider static allocation of a single message per slot: there is no interference between the messages. If a message misses its slot, it must then wait for the next slot which is associated with it. Additionally, the access delay of a message to the bus is then the maximum time which separates two consecutive slots which are assigned to the message m (Tm_{max}). In this configuration, the worst-case arrival time of a message m is given by the formula:

$$a_m = \underbrace{Tm_{max}}_{access\,delay} + \underbrace{X_m}_{propagation\,delay}$$

Additionally, the suitability of a preemptive policy for scheduling tasks in hard real time systems has been shown. In fact, scheduling tasks in systems which communicate on a TTP network has often been done using a non-preemptive policy with static priorities (Kopetz, 1997).

Approach Based on Fixed Priorities in Communications

In the event-triggered class of protocols, access to the communication medium is based on the priorities of the messages. Assignment of these priorities can be based on one of the approaches which have been met previously in the case of single-processor scheduling: FIFO, FP or EDF. In practice, the best-known protocol based on fixed priorities is the CAN protocol. In a system which is implemented with this, each message has a unique priority, and the message with the highest priority accesses the network. In *Calculating Controller Area Network(CAN) message response time* (Tindell, Burns and Wellings, (1995)), the authors analyse the WCRT of a message in a CAN network by taking into consideration two kinds of delay: delay in access to the medium and transmission delay. First they define the propagation time of a frame m of n bytes ($n \leq 8$) in a CAN network, by the following formula:

$$C_m = \left(\left\lfloor \frac{34+n}{5} \right\rfloor + 47 + 8n \right) \times \tau_{bit}$$

The CAN network is a resource which cannot be shared. Consequently, scheduling the network is related to scheduling tasks in a single-processor, non-preemptive context. In a holistic approach, the WCRT of a message m can be defined as:

$$R_m = t_m + C_m$$

where

t_m is the access delay to the medium, and C_m is the communication delay.

$$t_m = B_m \underbrace{\sum_{\forall j \in hp(m)} \left\lceil \frac{t_m + J_j + \tau_{bit}}{T_j} \right\rceil \times C_j}_{Interference}$$

where:

B_m is the blocking factor of m; the maximum wait time before accessing the bus.
J_j is the access jitter to the medium relative to the message j.
τ_{bit} is the transmission time of a bit.
T_j and C_j are respectively the period and duration of execution of the message j.

The authors also extend their model to take account of the error management mechanism which the protocol supports, and propose a much finer bound. In fact, in a CAN network, both the sending node and the receiving node of a message can detect an error. This error is signalled to the source node of the message, which consequently retransmits it. In the worst-case scenario, recovery from an error can necessitate transmission of a maximum of 29 bits, in addition to the message in question. However, despite the existence of a bound on the transmission time in a CAN network, it is not considered sufficiently deterministic for hard real time applications, since it is impossible to predict precisely the instant of transmission of a message, because of the technique for access to the medium.

As we have seen throughout this book, the FlexRay protocol is based on a hybrid approach, which makes it possible to combine determinism and flexibility. Before concluding this appendix, we will therefore present and discuss some existing state-of-the-art work concerning analysis of the time performance of FlexRay.

Scheduling Communications in FlexRay

A general recommendation for scheduling FlexRay messages is to dedicate the ST segment to hard real time messages and the DYN segment to soft real time messages. However, it is not always possible to follow this technique, for several reasons.

In fact, given that the size of the two segments is fixed in the system design phase, it is not always certain that sufficient unused slots remain to allow future expansion and development of the system. In this context, the properties of the flexible time division multiple access (FTDMA) mechanism, on which the DYN segment of the FlexRay protocol is based, were studied several years ago (Böke, 2003). The performance of this policy for access to the medium has been analysed, and conclusions have been drawn about its suitability for transferring messages with strict time constraints. Additionally, when analysing the Byteflight protocol, which is based on the same technique for access to the medium as the DYN segment of FlexRay, it is recommended that low identifiers should be allocated to hard real time messages, to guarantee compliance with their time constraints. However, this analysis was restricted to a virtually TDMA transmission scenario, which would cause the DYN segment to behave like the ST segment, and thus lose the flexibility provided by FTDMA.

In *Timing Analysis of the FlexRay Communication Protocol* (Pop et al., 2006), the authors propose a method of evaluating the WCRT in a FlexRay network, in the context of a holistic analysis. They define the WCRT R_m of a message m which is scheduled in the DYN segment as:

$$R_m = \tau_m + BusCycles_m(t) \times T_{bus} + w'_m(t) + C_m$$

In this equation:

τ_m is the waiting delay of a message during a communication cycle in the case that it is generated by the emitting task after its dedicated output slot;

$BusCycles_m(t)$ is the number of communication cycles in which the message m waits to be sent because of messages with higher priority than it;

w'_m is the delay after the start of the transmission cycle, from m until its slot;

C_m is the communication time of the message m:

$$C_m = Frame\_size(m)/bus\_speed$$

The authors propose a method of finding an optimal solution for $BusCycles_m(t)$ and w'_m. The optimal solution is obtained by modelling the problem in a linear problem in integers, which maximises the number of communication cycles in which the message cannot be sent. Its solution is obtained by the solver CPLEX 9.1.2. Once $BusCycles_m(t)$ has been obtained, they deduce w'_m, and identify scenarios in which a message m cannot be sent as being the consequence of:

- higher priority messages than m, so having a lower frame ID;
- messages from the same node and sharing the same slot as m (this configuration is not described in the protocol, but is purely application-dependent);
- the number of unused minislots before the transmission slot of the message m.

COMMENT

To our knowledge, all the work on scheduling messages in the DYN segment of the FlexRay protocol assumes that frame IDs have already been assigned, and consequently analyses the obtained performance of the protocol. Our approach will be different. In fact, depending on the application, a frequent requirement is to find a method of assigning identifiers which makes it possible to guarantee compliance with (at least some of) the time constraints of the system.

Policy of Assigning Priorities

A distributed system based on the FlexRay protocol is configured during the design phase. Once the nodes and messages to be transmitted are defined, the system architect must assign identifiers to the different messages, while considering the influence of the position of a message in the communication cycle on the response time, not only of the message in question but also of the other messages of the system. This strong dependence between the messages suggests a holistic analysis, to guarantee compliance with the time constraints of the system. However, the problem is not the same for the ST segment. In fact, the technique for access to the medium (generalised or global time division multiple access (GTDMA)) makes it possible to guarantee the predictability of the system.

More precisely, once the scheduling table has been established for the messages of the ST segment, it is certain that each message will be transmitted in the time window reserved for it (slot). The question is therefore about knowing how it is possible to guarantee compliance with the time constraints for messages which are scheduled in the DYN segment. Additionally, in *Bus Access Optimisation for FlexRay-based Distributed Embedded Systems* (Pop et al., 2007), the authors showed by experiments that increasing the size of the communication cycle increased the response time of the messages. Similarly, they came to the conclusion that a too short communication cycle would also degrade the performance of the system. They therefore proposed an optimal method of obtaining the network parameters.

In this appendix, we will disregard the ST segment, and consider the network parameters (size of communication cycle, sizes of ST and DYN segments) as already fixed, which is most often the case in industry, where a configuration which will be applied to a whole platform is defined. Additionally, we will consider that the size of the macrotick has been fixed so that communication jitter is taken into account.

To answer the problem, in this appendix we will use schedulers which are defined in the case of single-processor systems as a basis from which to identify the most suitable algorithm which will enable us to assign identifiers to the messages to be scheduled in the DYN segment. We will thus model our system to identify the class of scheduling problem. Consequently, we will be able to study the algorithm which we will use. Finally, we will consider a set of messages which must be scheduled, and it will be necessary to evaluate the quality of our approach by simulations, the aim being to comply with the time constraints concerning the messages.

Class of Scheduling Problem

After all of these pages, the only purpose of which was to document the subject for you in detail and to have a common vocabulary to elucidate these problems, here at last is your reward! To help you to solve the problem of scheduling the DYN segment, here are the few initial (realistic) starting assumptions which we have used:

- we will ignore the possibility of using two channels for communication (as of today, the conventional implementations of FlexRay are not (yet) of X-by-Wire type);
- we will consider:
  - an architecture of nodes which are interconnected by a FlexRay channel;
  - that a set of messages that must be scheduled in the DYN segment of the communication cycle is associated with each node;
  - that the allocation of messages to slots is static (defined once and for all), and that two different messages at a node will have two different slots, as defined in the specifications (no multiplexing at slot level);
  - that the arrival law of messages of the DYN segment (supply of the signal by the producing task) is sporadic, which is very broadly the case of applications today;
  - that time constraints apply to the deadlines, and that there is no a priori relationship between the periods and deadlines of messages;
  - that the deadlines concerning the termination of messages are derived from the time constraints of the tasks that receive the messages.

In a system which communicates via a FlexRay network, a message which has begun transmission cannot be interrupted to send another message, even one of higher priority. Additionally, transmission of a message is based on its identifier, which is fixed from the design phase and does not change while the system is in operation. Scheduling messages in the DYN segment of this protocol can therefore be treated similarly to scheduling in a single-processor system where the scheduler is non-preemptive and with fixed priorities.

The response time R_m of a message m transmitted in the DYN segment of the FlexRay protocol is therefore given by:

$$R_m(t) = \underbrace{\max_{t \in S}}(w_m(t) + C_m + J_m - t)$$

where $S = \{kT_m, \ k = 0 \ldots K, \ k \in \aleph\}$, where K is such that

$$w_m(KT_m) + C_m - J_m \leq (K+1)T_m$$

where

$w_m(t)$ is the delay caused by transmission of higher priority messages than m;
C_m is the communication time of the message m:

$$C_m = Frame\_size(m)/bus\_speed$$

J_m is the jitter which models the dependency between the scheduling of tasks and the communication of messages.

We put ourselves in the worst-case scenario, where we will suppose that transmitting a ready message can only be delayed by messages which have lower frame IDs than the message m in question, or if it is ready just after the slot which is reserved for it. (In our model, we will not consider the network parameter *pLatestTx* as being a cause of delay in sending a message on the network.) The arrival law of messages is periodic (maximum load of network). It can therefore be specified that:

$$w_m(t) = \left\lfloor \frac{t}{T_m} \right\rfloor \times C_m + \sum_{j \in hp(m)} \left(\left\lfloor \frac{J_j + w_m(t)}{T_j} \right\rfloor + 1 \right) \times C_j$$

Scheduling Algorithm

Taking into consideration the characteristics of our model of tasks (messages), the most appropriate algorithm is of FP type (see above). This algorithm is also valid in a non-preemptive context. We will now present the basis of the algorithm for assigning priorities which it has proposed:

$m = \{m_1 \ \ldots \ m_n\}$: a set of messages;
prio $\leftarrow$ n: *integer*; j: integer;
failed $\leftarrow$ *false*: Boolean;
while (m $\neq \phi$) do
   j = test-if-feasible (m, prio);
   if (j $\neq$ 0 AND failed = false) then
      assign-priority (j, prio);
      m = m $-$ {m$_j$};
      prio $\leftarrow$ *prio* $-$ 1
   else

```
        failed = true;
    endif
end
```

This assignment algorithm functions iteratively from the lowest priority to the highest. The 'test-if-feasible' function is based on calculating the WCRT which we have presented in the sections above. This function returns the first schedulable message m_j, or 0 if no message is schedulable in the current set of messages.

> **MATHEMATICAL REMINDER**
>
> $\forall x \in \mathbb{R}$, $\lfloor x \rfloor$ designates x rounded down to an integer, and $\lfloor x \rfloor$ designates x rounded up to an integer.
>
> Let τ be a set of tasks, and $x \in \mathbb{R}$. By convention, $\max^*{}_\tau(x)$ designates the maximum value of the parameter x in τ if $\tau \neq \emptyset$.

Conclusion

This non-limiting access and the associated problems have been presented purely with the aim of introducing you to these problems and confronting you with the general problems of scheduling messages which are transmitted in the DYN segment. Of course, to be purist, it will be necessary to extend these explanations and calculations to the use of FlexRay operating with two communication channels. In the case that redundancy is not wanted, it is sufficient to specify, for each message, the set of messages of the same channel, and in the case of redundant systems for secure solutions, it will be necessary to take account of the waiting delays of a message before being processed in the voting mechanisms.

To conclude on this subject, it is good that you should know that in the past 5–10 years, numerous very precise theses have been written on these subjects, by researchers on the teaching staff of engineering schools, by PhD students and by engineers on the permanent staff of automotive or equipment manufacturers. Yes, we have their names!

Part C

The FlexRay
Physical Layer

This third part is about everything that concerns the physical layer of the network, directly or indirectly, and more particularly its effects and repercussions on the structure of the FlexRay protocol.

This layer is one of the most complex and difficult to grasp. The properties and performance of the physical layer and the choice of the structural design of the protocol are very intimately linked to, on the one hand, the high bit rate of 10 Mbit/s, the principle of time division multiple access (TDMA) which is used for access to the medium, an architectural and distributed intelligence philosophy with shared synchronisation, and, on the other hand, a wish for redundancy to comply with a high level of security in operation.

As in the case of CAN, the medium of the physical layer as described in the official specifications of the FlexRay 2.1 protocol is not explicitly defined, and leaves the door open to several implementation possibilities such as wired media of differential pair type or optical fibres. However, to have a specific base and fix the first implementations, version 2.1 of the FlexRay physical layer specification describes in detail only the case of a wired medium with differential pairs.

It should be noted that a strongly distinctive feature of FlexRay relative to CAN is the fact that a node must be capable of supporting two totally independent physical layer channels simultaneously, defined as 'channel A' and 'channel B'. As we will show later, this has two purposes, the first being the possibility of communicating at faster speeds during phases when the network is functioning well, the second being the possibility of providing functional and/or physical redundancy for data transmission in the case of an incident occurring on one of the two transmission channels, and thus increasing the 'fault-tolerant' feature of the system.

Additionally, given that the ultimate purpose of a well-designed communication system is that it functions correctly – that is with the lowest possible bit error rate (BER) – a fundamental purpose of this part of the book is to examine the various aspects of the topology of a FlexRay network and their effects on the quality of the transmitted and received signal. To do that, we will examine in detail all the parameters which are involved in the routing of the signal, from its creation to the validation of its received binary value. Thus, to explore as fully as possible the problems associated with the physical layer, we have decided to divide this part into several chapters, and to present them deliberately in the sequence corresponding scrupulously to that of the propagation of the signal from one node to another, and at each level of the link, we will describe the functions and effects related to it. In the course of the chapters, you will therefore find:

- Creation and emission of the signal
  - logical and then electrical creation of the signal, its encoding, its form and its amplitude
  - line control stages
  - electromagnetic compatibility (EMC) and electrostatic discharge (ESD) protection of the line control stage
- Transport of the signal
  - medium
  - usable types of medium
  - theoretical propagation of the signal and its effects on this medium
- Topologies of the medium
  - usable topologies
    * single channel/dual channel
    * single channel
    * linear
    * linear + stub
    * star, active star, hybrid, dual channel and redundancy
  - consequences of the topologies on the integrity of the signal
    * rise time/fall time
    * asymmetry
    * ringing, crosstalk
    * influence of the components arranged on the medium
    * drivers, repeaters and stars
    * asymmetry of the output and input stages
    * starting the signal
    * return time
    * EMC filtering
    * ESD protection
- Reception
  - reception stages and their performance
  - interference and its effects
  - bit validation processing
- BER – modelling the link
  - modelling and evaluating the consequences
    * Monte Carlo methods

   * concluding the modelling and taking account of the BER
   * eye diagram
  – broad outline of recommendations for use
  – examples of topologies with comments
- Components
  – electronic components which are designed for practical implementation of the physical layer

8

Creation and Transmission (Tx) of the FlexRay Signal

The purpose of this chapter is to describe and comment on how the signal to be transmitted at a local node is generated and transmitted. It includes the generation of the signal within the CC and then its passage through the line control stage.

8.1 Creation of the Signal

The electrical signal (binary elements – bits, frames, and so on) to be transmitted is created by the CC, which is either completely within a specific integrated circuit outside an application 'host' microcontroller, or integrated directly into it. Its purpose is to produce successions of bits of nominal duration 100 ns. For fuller information, the internal architecture of the CC is given in Chapter 18 about components.

Let us begin with a physical description of the binary logical and electrical signal which the medium will carry sooner or later. In principle, it is characterised by three theoretical aspects: its encoding, its speed and its physical representation.

8.1.1 Bit Encoding

The term 'bit encoding' describes the theoretical representation of the logical bits '1' and '0' which form the transmitted data. In the case of FlexRay, the bit encoding which is used is of 'non return to zero' type, meaning that once the value of the physical signal has been established, it does not change throughout the duration of the bit (see Figure 8.1).

As a reminder, as we indicated about encoding communication frames, the succession of bits is organised in bytes of 10 bits 'NRZ 8N1'; that is, 8 NRZ-encoded bits framed by a 'START' bit and a 'STOP' bit.

8.1.2 Bit Rate

The original FlexRay specification indicates that the only nominal value of the gross bit rate is 10 Mbit/s, that is a bit duration of 100 ns – enabling us to predict, as we will

---

FlexRay and its Applications: Real Time Multiplexed Network, First Edition. Dominique Paret.
© 2012 John Wiley & Sons, Ltd. Published 2012 by John Wiley & Sons, Ltd.

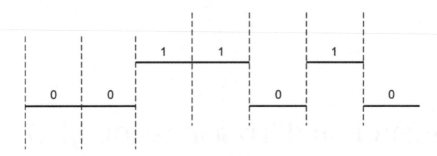

Figure 8.1   'Non return to zero' encoding

show in the following paragraphs, that the slightest little delay times, signal propagation times and network topology times will have a big effect on the quality and integrity of the transported bit.

It should be noted that although the speed is specified as 10 Mbit/s, the original document indicates – for information – the values that can be used for speeds other than 10 Mbit/s. So you want to know what happens now? We will return to this sensitive point at the end of the book.

So much for the theory of the creation of the signal. At this level, a long alternating sequence of logical '1–0–1–0–1–0, and so on' each of duration 100 ns, represents on paper a square signal of 5 Mbit/s with a strictly symmetrical 50/50 duty cycle.

8.1.3   The Communication Controller (CC)

Reality is already a little different, because for numerous reasons, the transistors of the integrated circuit of the CC can cause a slight asymmetry/dispersion/tolerance/degradation by a few nanoseconds of this duty cycle. The FlexRay specification indicates that this must not exceed 2% or 2 ns.

> **NOTE**
>
> We know that 2 ns may make you smile, since it represents only 2% of the duration of the signal, but as things are now, when your boss gives you a 2% pay rise, you thank him all the same – so 2% isn't as negligible as all that!

Once the bit encoding has been determined, to implement an actual appropriate line driver stage, its physical representation must be defined. Let us now go on to examine the signal which must be applied to the communication line via the line driver.

8.2   Physical Representation of Bits

Let us begin with the theoretical physical result that the line driver must produce.

In principle, like CAN, FlexRay is capable of supporting different physical representations of the bit. Since it was necessary to start with something specific, with known

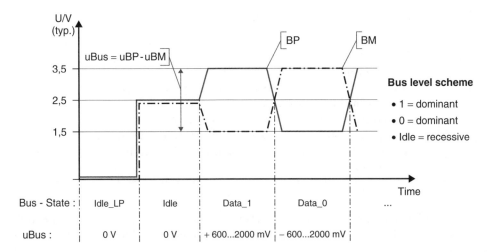

Figure 8.2 Physical representation of bits in the context of a medium of 'differential wire pair' type

properties and costs associated with the intended markets, the specification of the physical layer of FlexRay describes the physical representation of bits in the context of a medium of 'differential wire pair' type. And yes, the electrical signal is transmitted in 'differential mode', as in CAN, but, as indicated in Figure 8.2, differs from it in that the differential electrical levels alternate (change sign) to represent the values '1' and '0', and are consequently both represented by dominant states.

Surprising, isn't it?

Note that there is no representation of recessive binary logical values, that the presence of a recessive level is reserved for the 'idle' mode of the medium and that there is also a fourth level (two dominant, one recessive and this one) for 'low power down' mode.

NOTE

Unlike CAN, in FlexRay there will therefore be no time lag during which (for example arbitration zone/field) a dominant value will 'overwrite' a recessive value. In fact, because of the principle of the time slot used in TDMA (in the static segment) and flexible time division multiple access (FTDMA) (in the dynamic segment), there can be no conflict for access to the medium, each of the participants (and its associated frame) having its very precise time slot or minislot.

8.2.1 Differential Voltage on the Medium

By definition, the voltage between the two wires of the differential pair forming the medium is called the differential voltage (V_{Bus}):

$$V_{Bus} = (V_{BP} - V_{BM})$$

each of these two values being respectively measured relative to the reference potential of the system (earth, '0 V').

Taking account of these assumptions, the four states below are defined:

Data_1	$V_{Bus} = (V_{BP} - V_{BM})$ Positive
Data_0	$V_{Bus} = (V_{BP} - V_{BM})$ Negative
Idle	No data present on the network, but at least one node in the cluster which is not in 'low power mode'
	BP and BM are both at the Idle voltage level
Idle_LP	Low power modes (Sleep, GotoSleep, Standby, and so on)
	BP (bus positive) and BM (bus minus) are both earthed with pull-down resistors

It should be noted that the differential value of the voltage $V_{BP} - V_{BM}$ is of the order of 700 mV, so that the signals on the network contribute very little to electromagnetic radiation.

8.3   Line Driver 'Tx'

To be able to actually control the physical layer, sooner or later it is necessary to have, between the output of the protocol manager and the input to the medium, electronic elements pompously called 'line drivers' or 'transceivers (transmitters–receivers)', to control and interpret the changes of electrical level which are present on the communication line. They have transmission output stages 'Tx' which enable them to transmit the signal, and reception input stages 'Rx' which enable them to receive the incoming signal. Let us now examine the electrical parameters associated with the 'Tx' part.

8.3.1   Rise Time/Fall Time

The output stages of the line driver components 'Tx' generally consist of symmetrical (or almost!) power stages of so-called 'push–pull' type (see Figure 8.3). Unfortunately, in real life, it's always the 'almost' that makes the difference!

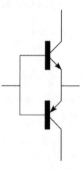

Figure 8.3   'Push–pull' type

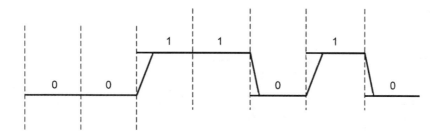

Figure 8.4   Integrity faults between the transmitted signal and the received signal

In fact, the often unavoidable asymmetry (even slight) of the physical implementation of the output stages (different output impedances between the high and low states of the push–pull stages) of the line drivers (and/or active stars – see below) with the same interference capacitance values on their output pins can cause asymmetries and significantly different values of the rise and fall times of signals which must be propagated on the network, and thus cause integrity faults between the transmitted signal and the received signal (see Figure 8.4). Additionally, these asymmetries can vary and evolve as a function of variations of temperature and power supply voltages, but in general they are considered to be constant (not to vary) while a frame is being received. For this purpose, the FlexRay specification indicates that these asymmetries are tolerable if they are limited to 4 ns for the transmitter (as preliminary information, 5 ns for the receiver and 8 ns for passing through an active star).

8.3.2   Impedance Matching

To ensure the optimal transfer of power, and to minimise reflections and other interfering phenomena at the transmission stage Tx (for example 'ringing', see below), the line must be closed on a resistive load, the value of which equals its characteristic impedance. In the case that the terminating charges of the network are not well matched (mismatched impedances in inductive or capacitive real and/or complex values) or not actually shared symmetrically relative to earth, this can cause the appearance of reflected waves, the presence of coefficients of reflection and stationary wave ratios, and cause what are called 'ringing' phenomena.

8.3.3   'Ringing'

Because of an impedance mismatch as described above, and/or the interfering introduction of reactive, inductive or capacitive (or of course both) components, at transitions of electrical signals corresponding to logical '1' to '0' (and vice versa) more or less damped overshoots (or undershoots) can appear, and be added to the original signals. As indicated in Figure 8.5, these overshoots participate in the modification of rise and fall times of the signal, and when they are detected cause asymmetrical distortion of the propagated signal.

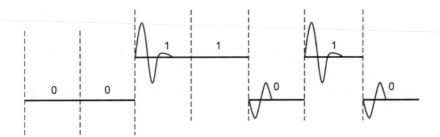

Figure 8.5   Overshoots (or undershoots)

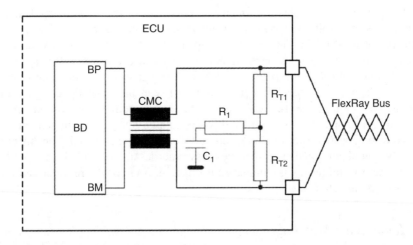

Name	Description	Typ	Unit
R_{T1}	Resistor of split termination	$\dfrac{Z_0}{2}\,4$	Ω
	Tolerance	1	%
R_{T2}	Resistor of split termination	$\dfrac{Z_0}{2}\,4$	Ω
	Tolerance	1	%
R_1	Resistor	5	Ω
	Tolerance	1	%
C_1	Ceramic capacitor	4.7	nF
	Tolerance	10	%

Figure 8.6   EMC filtering

8.3.4 EMC Filtering

To reduce the problems associated with electromagnetic interference (EMC), and to comply with the radio frequency (RF) pollution standards in force (for emission of radiation – the ETSI EN 300-220/330/440 families of standards), the rapid rise and fall times must be carefully controlled at the level of the signals which are transmitted on the communication line. In parallel (see the section about reception input stages for more detail), it is also useful to guard against external RF interference (immunity). With this double purpose, it is often useful to have, as close as possible to the output lines of the line drivers, a second order low-pass filter, implemented using a double symmetrical coil (generally called a 'common mode' coil) and capacitors which are placed on the two elements forming the differential pair which is used for the link. To be effective, the double coil must have, on the one hand, very low stray/leakage inductance and, on the other hand, a negligible hysteresis phenomenon (see Figure 8.6).

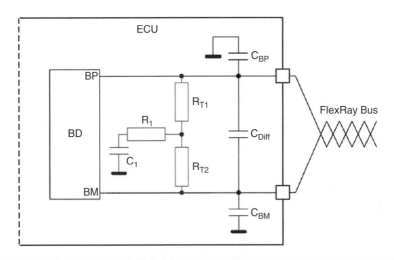

Name	Description	Typ	Unit
C_{BP}	Capacitance of BP to GND	47	pF
	Tolerance ; NP0 dielectricum	5	%
C_{BM}	Capacitance of BM to GND	47	pF
	Tolerance ; NP0 dielectricum	5	%
C_{Diff}	ECU's differential input capacitance	39	pF
	Tolerance ; NP0 dielectricum	5	%

Figure 8.7 ESD protection

The symmetries and complementarities of two windings of the so-called common mode coil should be such that they can degrade the asymmetry of the signal by only $\pm$ 0.25 ns at maximum.

8.3.5   Electrostatic Discharge (ESD) Protection

Obviously, it is also necessary to ensure that the components which are intended to protect the communication lines (protective diodes, voltage-dependent resistors (VDRs), capacitors, and so on), and which are arranged on lines or line terminations against classic ESDs (pulses of 'x' kV, positive, negative, in 'human model' and 'machine model') are well balanced.

The 'bible' for conformity tests on the physical layer of FlexRay (which is a veritable 830-page mine of information, which sooner or later you will be obliged to read to be certain of conforming to it ...), the document *FlexRay Physical Layer Conformance Test Specification*, indicates the diagram of Figure 8.7 as the reference testbed for ESD measurements.

9

Medium, Topology and Transport of the FlexRay Signal

This chapter is about everything to do with transporting the signal between two participants of the network. It is divided into two large parts. The first part deals separately with the medium, to define its qualities, performance and limitations. The second part concerns, more particularly, on the one hand the topological aspects and possibilities which FlexRay offers to the structure of a network, and on the other hand the electrical and functional consequences of these possibilities.

9.1 Medium

In principle, like CAN, the FlexRay protocol is in such a form that when its physical layer is put into concrete form, it is capable of supporting different media which, for example, are implemented using wire links (for example differential pairs), optical fibres and RF waves. Since it was necessary to start with something concrete, with known properties and costs associated with the intended new market, while waiting for what follows, the FlexRay specification describes the physical representation of communication signals which are implemented in the case of a medium of wired differential pair type.

Among these differential pairs, twisted pairs are always recommended, and as usual the screened aspect is no longer optional in practice. The conformity specification cited above gives the numbers in Figure 9.1 for the characteristics of the screen.

9.1.1 General

During a communication, several things happen:

- starting communication and starting the components of the network;
- then continuation of the flow of data forming the communication;
- and finally, stopping communication.

This list may seem to you puerile or quaint, but concerning FlexRay it hides important details.

FlexRay and its Applications: Real Time Multiplexed Network, First Edition. Dominique Paret.
© 2012 John Wiley & Sons, Ltd. Published 2012 by John Wiley & Sons, Ltd.

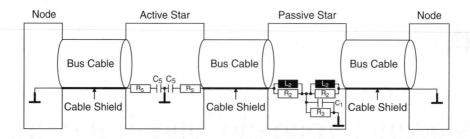

Name	Description	Typ	Unit
R_S	Damping resistance	1000	Ω
	Tolerance	1	%
C_S	Capacitance	470	nF
	Tolerance	10	%
L_2, R_2, R_3 and C_1	Components of the passive star, see chapter 2.4.7.		

Figure 9.1 Shielded cable

At the start of a communication, certain components take a certain amount of time to start, and consequently can truncate part of the message being sent. For more details, we refer you to the specific section about this subject at the end of the chapter.

Throughout the communication phase, obviously at the start (but sometimes masked by the truncating effect mentioned above) but principally observable while data are being transmitted, the effect of propagation of the signal on the medium occurs. We will now examine it in detail.

9.1.2   Conventional Propagation of the Signal on the Network

Whatever media and topologies are used to implement the network (point to point, linear bus, with passive stars, active stars, repeaters, and so on), the signal is propagated along lines formed by the network, and arrives at its destination a few moments after it leaves its source. This delay, which is due to the propagation of the signal on the medium, is directly linked to its propagation speed on the medium and to the length of the network section on which it is propagated.

9.1.2.1   Propagation Distance

If we ignore the more subtle concepts of the beautiful theories of Lorentz, Langevin and Einstein, as everyone knows the concepts of distance and time are linked through speed of motion: $l = v \times t \rightarrow t = l/v$.

In our case, the three parameters 'distance/network length', 'propagation speed of signal' and 'propagation time' are therefore linked to each other. Let's explain quickly what is hidden behind each of these parameters!

9.1.2.2 Propagation Speed

In the case of FlexRay, if we mention the parameter of speed of propagation of the signal on the medium on which it is propagated, several cases can occur in the same application.

In the Presence of a Single Medium Type (in Mono or Dual Channel Mode)
Like any speed, the value of the 'propagation speed' parameter of the medium which is used is expressed in metres per second. As a reminder, the speed of an electromagnetic wave in a vacuum or air is approximately of the order of $300\,000$ km/s $= 3 \times 10^8$ m/s or 0.3 m/ns.

For information, an example is given in the official specification of the physical layer of FlexRay, indicating not the propagation speed but its inverse – value of the delay/propagation time per metre (propagation delay) – of 10 ns/m for a wire line, suitable for loads of values between 80 and 110 Ω. The performance of the medium considered by FlexRay therefore corresponds to a propagation speed of 0.1 m/ns or $100\,000$ km/s. It should be noted that this value is quite pessimistic, since most conventional wire media have propagation speeds of the order of $200\,000$ km/s.

In the Presence of Two Different Media on the Same Network
Given that FlexRay can support two different types of medium equally well (wire and optical), and that they can be used simultaneously (one on one transmission channel, the other on the other, in the case of dual-channel applications), other things being equal (topologies, distances, clocks, and so on), it will therefore be necessary to take account of propagation time differences due to two different types of media. As the equation above has reminded us, through the propagation speed of the signal on the medium being used, the lengths (or distances) of the different parts of the network and between nodes therefore have a direct effect on the 'Time' or 'Global Time' parameter mentioned in earlier chapters.

9.1.3   Total Distances Used or Wanted on the Network or between Nodes

Obviously, it is always possible to assign a maximum value to the 'distance' parameter by examining the nodes at the extremities of a network, which of course must be taken into account, but beyond all that, the most complex problems are those associated with the structural inhomogeneity of the distances (and therefore of the respective propagation times) between the different nodes on the network as a function of the topologies which are used or desired – or otherwise imposed.

This implies that, sooner or later, it is necessary to be interested in the detail of the problems of '(time) symmetrisation' of networks and of estimating what are commonly called 'differential times'. Figure 9.2 gives an example.

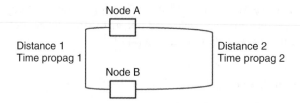

Figure 9.2   '(Time) symmetrisation' of networks and estimation of 'differential times'

9.2   Effects Linked to Propagation

9.2.1   Propagation Delay Time

The first effect is that of the delay time caused by propagation of the signal. We will mention two aspects:

- the theoretical aspect of this propagation, which on paper means that the signal is simply shifted in time, without being deformed and with its integrity unaffected. This is what we will describe in detail in Section 9.2.2;
- the reality is quite different, because very often non-linear, second order phenomena interfere with the beautiful theories of physics.

9.2.2   Symmetrical Effects

The theoretical effects of the propagation of a signal on a communication network involve three important things:

1. the delay time due to the propagation of the signal is due to the type of transport medium which is used and the propagation speed on it;
2. the fact that the propagation delay is not linked only to the type of transported signal (shape, duration), but to be precise that in our case the leading and trailing edges of the transported signals are delayed identically, which has no effect on its timing integrity and respects the initial symmetry of the duty cycle of the transported signal;
3. the duration of the initially transmitted bit is therefore identical when it arrives.

Summarising, once the signal has left, everyone is singing from the same hymn sheet, the delays of the leading and trailing edges are the same, and this delay is a simple shift in time of an initial phenomenon, and **has no effect on the timing integrity of the signal**, its duration, and so on.

In principle, in this case, a sequence of FlexRay bits 1 0 1 0 1 0 1 0, and so on at 10 Mbit/s, with no return to zero (NRZ) bit encoding, should be represented by a square signal of 5 Mbit/s with a 50/50 duty cycle (100 ns ON, 100 ns OFF).

Figure 9.3 illustrates this phenomenon, and indicates what is generally called the propagation delay.

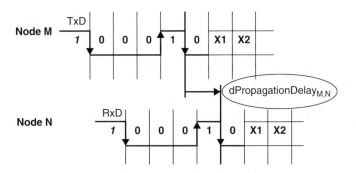

Figure 9.3   The propagation delay

IMPORTANT NOTES

For numerous other reasons which we will mention below, the maximum propagation time that FlexRay specifies must not exceed $2.5\,\mu s = 2500\,ns$. Because of a propagation delay of 10 ns/m, do not conclude hastily from the lines above that the point-to-point length of a network could be 2500 ns/10 ns/m = 250 m – because this is wrong, and is not how you should look at it. At the level of digital processing of the signal, whatever physical delays are encountered on the network (medium, line drivers, repeaters, stars, and so on) two systems must not be 'logically' distant in time by more than 25 bits; that is, $25 \times 100\,ns$.

The influence on timing of the use of active stars (see below in this chapter) can be considered as equivalent to that of a repeater, and so on, which obviously takes time to carry out its redirection, and this time must be added to that of the pure propagation of the signal. This is generally estimated at a value of the order of 200/400 ns. Also, frequently, the topologies used by the applications use two active stars per network, which immediately takes up about 400–1000 ns of the total propagation time which is available to the FlexRay protocol.

9.2.3   Reflection, Matching

Propagation on a communication 'line' means termination impedances, potential reflections, coefficients of reflection, matches, mismatches, ringing and other little treats of Smith's abacus type and others $(1 - \Gamma^2)$ here, there and everywhere. In short, the usual!

Given that these phenomena and parameters are strongly associated with the media that are used, the FlexRay specification indicates clearly that the use of wire links (cables, twisted pairs usually with electrical screens) is linked to their performance, by imposing characteristic impedance values between 80 and 100 Ω, the maximum propagation time being 10 ns/m and the maximum attenuation being 82 dB/km.

Additionally, in the case of use of two transmission channels (A and B), it is strongly recommended that the differences should be minimised, and the delays due to propagation of signals on the two channels should be balanced as far as possible – which is sometimes or often easier to say than to do!

9.3   Topologies and Consequences for Network Performance

As we have stated, this second part of this chapter deals with the topology which is applied to use of the FlexRay protocol. First let us remember that the etymology of the word 'topology' comes from the Greek 'topos', place and 'logos', study, and take it as meaning 'the study of properties of places'. And that's what we will do ...

Given that the ultimate aim of a communication system is that it functions correctly – that is with a known bit error rate (BER) which is the lowest possible (see next chapter) – in the course of this part of the chapter we will study the various aspects of the topology of a network and their effects on the quality of the transmitted and received signal. To do this, we will examine all the aspects involved in the routing of the signal, from its production to the validation of its binary value.

9.3.1   First, a Little Light on the Obscurity of the Vocabulary

Because of the numerous mechanical and geometrical forms of networks, their physical and electrical properties are very different. The forms and topologies of networks are very varied: bus, bus with stubs, star, ring, hybrid, and so on. Far too often, the name 'XXX bus' is attached to simple 'names of protocols', which is wrong and often contributes to abuses of language and broad confusions of the type of 'CAN bus' instead of 'CAN'. I can bear witness to this personally, since I have often been guilty of it – and to top it all, voluntarily! A fault confessed is half pardoned? So confess it twice, and you're completely pardoned! No! In fact, at that time – the 1990s – we should have spoken simply about 'CAN', as I should have done in numerous works which I wrote on the subject – but on the one hand to avoid creating confusion with CAN meaning 'convertisseur analogique numérique' (French for AD/DA converter), and on the other hand because the most suitable topology for good operation of this protocol is that of a bus, it was decided by some high technical and editorial authorities to call it the 'CAN bus' in France ... which is structurally wrong! In short, this time, in the case of FlexRay, since there is no reason of the same kind for removing the confusion, we will speak only of specific topologies which are applicable to FlexRay, whether they are those of buses, stars or whatever!

9.3.2   Effects and Consequences of the Topology of a Network on its Performance

Before entering specifically into the detail of the numerous topologies which can be applied to FlexRay networks, we will first go quickly through a few paragraphs to examine and comment in general on the principal effects of the topology of the network on its electrical performance. But first, as an introduction, a few important words concerning two radically different application philosophies.

- **Applications to topologies which are fixed once and for all:** In this type of application, after long reflection, the topology is chosen, validated and does not change again. As the lines below will show, this case is already not simple, but ideal! The typical case of this application field is, for example, the design of a unique vehicle model. This is often the case with a so-called 'top-range' model.
- **Applications to evolving topologies depending on potential options:** In this family of applications, the wish is no longer to develop a unique model, but an application platform which can support numerous variants or options involving both the number

of nodes which can be connected and the addition and/or removal of connection stubs. This is often the case with a 'mid-range' platform with multiple options. In short, this becomes very complicated, because it is necessary to imagine all the cases and sub-cases of the different topologies – and their consequences!

Now that you have been warned, let us go on to review all the network topology parameters which may be affected.

9.3.3   Distances between Two Elements

This is the first parameter that comes to mind. The distance between two nodes can be said to affect directly:

- the attenuation of the received signal because of line losses (ohmic resistance of the wires);
- the propagation time of the signal, linked to the properties of the medium (copper, aluminium, optical fibre, and so on), which, depending on what protocol is used or intended, can make it unusable in the desired application. As a reminder and to be very clear, it is fundamentally important that the sum of the 'there and back' propagation times within the duration of the acknowledgement bit limits structurally the operating distance and the use of CAN!
- the signal which, because of the interfering capacitances and inductances on the network, has every chance of being deformed, and consequently:
  - having an asymmetrical propagation time,
  - causing problems, in particular of:
    * the precise choice of the sampling point of the signal
    * line termination impedance mismatching,
    * creation of standing waves and standing wave ratios,
    * 'ringing' (overshoot and undershoot),
    * 'jitter' of the time position of the edges of the bit, by any of:
      o earth noise,
      o distant power supply noise,
      o crosstalk between adjacent wires,
    ... and thus causing lack of integrity of its value and a high BER.

9.3.4   Distances between Several Nodes

When 'n' participants are arranged on a network, the problem is identical to what we have just described for two nodes ... to the power of 'n'! In fact, the distances between the nodes are never strictly identical, and therefore adaptations and propagation delays are often not merely folklore.

Obviously, to solve these problems, it is possible to imagine arranging all the nodes according to a 'star' topology, all the branches of which would be of strictly identical length. It is sometimes nice to dream, because for many basic, mundane reasons that is practically never possible. Just for fun, Figure 9.4 shows this nice dream.

Passive Star

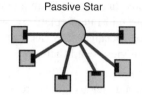

Figure 9.4   '(Time) symmetrisation' of networks and estimation of 'differential times'

9.3.5   Relationships between Topologies, Electromagnetic Compatibility (EMC) and Electrostatic Discharge (ESD)

There are always direct relationships between the various topologies which are used for constructing networks, the slopes, rise times and fall times of the transported signals, the phenomena of overshoot and undershoot ('ringing'), and therefore the devices which are used to provide ESD protection and counter EMC pollution.

In general – ignoring specific features of FlexRay – the loop which surrounds the extent of the problems associated with all these parameters takes a long time to grasp and resolving it usually results in numerous compromises to be carried out. If a layer which is specific to the protocol of FlexRay type is added, it is also necessary to take account of the symmetrical and asymmetrical delays, the quality and integrity of the signal, the 'truncation' phenomenon and other related little treats. In short, have fun!

9.3.6   Integrity of the Signal

The signal output by a node has a definite form which, in principle, must not be deformed to preserve its integrity. Unfortunately, numerous parameters can alter it.

9.3.6.1   Due to the Medium Itself

Let us list the principal reasons which cause asymmetries of rising and falling signals, and which may be due to the medium itself and how it is specifically implemented.

Even after taking care to choose and use a differential pair which is ideally symmetrical, when it is implemented in the application, for simple reasons which are completely outside the wishes of the network designer, the following may exist:

- asymmetrical capacitive coupling between the propagation lines of the network (the differential pair) and certain tracks and intermediate layers of multilayer printed circuits, earth planes, and so on;
- asymmetry of the capacitance values of simple wires and linking cables, and often between the pins of connectors;
- 'ringing' phenomena (overshoots or undershoots following signal transitions) between the pairs and other cables consisting of strands ('harness'), also causing reduction or

lengthening of slopes by reflections of the signals, producing asymmetries of the signals as explained above;
• crosstalk, and so on (see below in this chapter).

9.3.6.2 Due to the Topology and to Various Components which are Arranged on It

In the same way as the line driver stages which were described at length in the course of the previous sections, it often happens that the components that form and participate in the topological implementation of the physical layer (active stars, repeaters, and so on) deform, alter, truncate in the cumulant the propagated signals, and modify their rise and fall times.

For the curious, the numerous topologies which can be used with FlexRay and their respective qualities and performances are presented in a few sections. The only purpose of this section is to remind you that each of them – via the elements which are arranged on the network – will bring (or not), depending on its complexity, its batch of asymmetrical delays to the work which you will compose, and you must take account of them ... and above all, later you will be judged entirely responsible for them!

Summarising, as far as topologies are concerned, the biggest actuators of asymmetrical delays are the active stars (whether or not they include a local intelligence, with or without a microcontroller) and their successive cascading, and/or the repeaters – which are actually nothing but disguised active stars.

> **IMPORTANT NOTE**
>
> Don't forget that it is often necessary to have, on each branch of the star, an impedance matching load, an EMC common mode filtering coil and a device for protection from ESD.

9.4 Single-Channel, Dual-Channel and Multi-Channel Communication Topologies

The nodes of a network can be linked to each other in different ways, either by a single transmission channel or by two or more channels.

9.4.1 Topology of Systems of Single-Channel Type

An example of this generic type of topology is given in Figure 9.5.

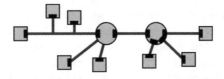

Figure 9.5 Topology of systems of single-channel type

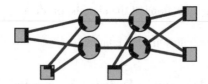

Figure 9.6   Topology of systems of dual-channel type

Simply by its form, the topology with a single communication channel:

- reduces the costs generated by the cabling and harness;
- avoids having a spatially nearby second channel;
- is based on long application experience in the automotive field;
- joins known costs in the car.

9.4.2   Topology of Systems of Dual-Channel Type

An example of this generic type of topology is given in Figure 9.6.

By definition, FlexRay must be capable of supporting the use of two transmission channels, A and B. It should be noted that the FlexRay specification says nothing about their respective uses. That leaves open a multitude of possible applications, with the same topology.

9.4.2.1   Systems without Redundancy

These two channels can be used to connect the nodes on the network, partly or fully, for transporting information with no redundant value between them, for example:

- either for transmitting data which are strictly unrelated (neither complementary nor redundant) between them, and thus apparently increasing the overall bit rate of the network;
- or, in normal operation, to allow the use of two channels for transmitting complementary (but not specifically redundant) information at an even higher rate, for example a gross bit rate of 10 Mbit/s on each channel; that is, an overall bit rate of 20 Mbit/s.

9.4.2.2   Systems with Redundancy

After this taster about high-speed, non-redundant architectures, using one or two communication channels, let's slip in a few words about redundancy in systems.

Controlling the actions of the brakes via a wire link called '*Brake by Wire*' is an example of the type of system in which it is preferable to ensure some redundancy in operation! So goodbye to master cylinders which can leak, pipes of any type, special liquids, and so on, and welcome to fine electric motors with their associated local electronics mounted level with the brake callipers, and acting to control the worm screws, the purpose of which is to press the struts against the discs (see Figure 9.7, from Siemens-Continental).

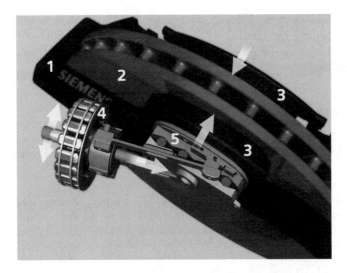

Figure 9.7   Brake-by-Wire example. Reprinted by permission of Siemens – Continental

We'll spare you the details, but that certainly has to work every time! Two precautions are better than one, but that's no problem, let's double the communication networks:

- either to transport, on each of the channels, strictly the same information twice, to have redundancy in the strict sense;
- or, in case one of the two transmission channels fails, to support concurrent, redundant data transmission on the other and to have a known fallback position.

9.4.2.3   Systems with Reserves/Application Options

In the same way, when the network is designed, the second transmission channel is often initially put entirely or partly in reserve, to allow for implementation of options, or for future improvements to the system, or for other or new product ranges which are included in the company's 'Reuse' and 'Time to market' strategy and policy.

A last little mischievous comment to conclude these paragraphs, which are specifically dedicated to topologies using two communication channels: you should bear in mind, and it's a good bet, that duplicating a link between two nodes by creating two different communication routes (for reasons of mechanical and electrical redundancy, and so on) simultaneously results in creating two routes of different lengths, and thus forming a (mini) ring structure and the associated worries!

9.5   The FlexRay Topologies

The very high bit rate leads us to be very careful concerning the possible topology(ies) of the network. In fact, propagation time, rise time, fall time, radiation, information redundancy, and so on will be the keywords of our and your reflections.

Having said that, beyond structures with single and dual communication channels, let us now examine the large families of possible links between participants of a network, and let us begin with the simplest possible link between two nodes; that is, the so-called 'point-to-point' link.

9.5.1   Point-to-Point Link

Taking account of the parameters mentioned above (propagation time, and so on), the FlexRay specification indicates that a length of 24 m should not be exceeded. In the case of a wire link which is implemented using a differential pair, this can be designed using bidirectional line drivers, as shown in Figure 9.8.

It should be noted that it will be necessary to arrange, at each end of the line, a termination resistor (for line impedance matching), to avoid the occurrence of standing reflected waves which degrade the signal in shape and received power, in the same way as must be done for CAN.

9.5.2   Link Using a Passive Linear Bus

Let us extend the link of point-to-point type into a linear bus topology.

This is the most economical linking topology, which comes to mind first for uniting different nodes – on paper (see Figure 9.9).

Let's return to this point. 'Linear bus topology' must be understood as a bus topology with terminations which are actually and concretely on the bus, and not at even a slight distance from it. Sometimes this is not physically simple to do (difficult cable raceways, mechanical positioning of complex modules), which makes this topology often conceivable only on paper!

In this case of topology, the FlexRay 2.1 specification indicates some restrictions, see Figure 9.10.

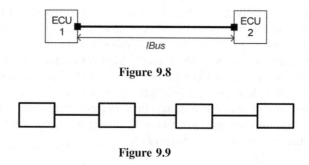

Figure 9.8

Figure 9.9

Name	Description	Min	Max	Unit
IBus	Lengthofapoint-to-pointconnection		24	m

Figure 9.10

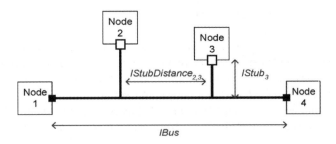

Figure 9.11

It should also be noted here that only matched end-of-line terminations are adequate for good operation. Given that, because of a low potential reflected wave ratio as a result of this network configuration, this topology has by far the best performance concerning, on the one hand, the quality and integrity of the signal, and, on the other hand, solving problems related to EMC.

9.5.3 Link Using a Passive Linear Bus with Stubs

This linking topology is very widespread, since it is also part of the most economical solutions for uniting the various nodes of a network. Sadly, all the nodes are not arranged directly and strictly on the (passive – no active component is arranged on the transmission channel) bus, but they are often attached to it using shorter or longer 'stubs', as shown in Figure 9.11.

Because of the content and large harmonic spectrum associated with the high bit rate of FlexRay, these stubs behave like branches of a communication line, and according to the relative wavelengths of the harmonics of the transmitted signal, form what are usually called 'stubs', in the same way as those that are found in UHF (ultra high frequency) and SHF (super high frequency) when lines are studied. Consequently, antinodes and wave nodes can be produced on the bus, cancelling or amplifying locally the voltages at the points where the stubs are attached to it. To avoid these problems, and taking account of its own specific features, the FlexRay specification also indicates some particular values which must not be exceeded:

Maximum distance between two nodes in a system	24 m
Distance between two splices of the network	150 mm
Maximum number of nodes with stubs	22

As a reminder, the FlexRay bit rate is 10 Mbit/s, with bit encoding of NRZ type, or an equivalent square signal (1 0 1 0 1 0 1 0 1 0 ...) of maximum frequency 5 MHz, including only odd harmonics. If it is desirable to satisfy correctly the eye diagram presented in Chapter 11, it is necessary to be able to pass cleanly at least harmonic 30 – that is at least 150 MHz. As a reminder, a 100 MHz wave has a wavelength λ of 3 m, so that $\lambda/2$ is 1.5 m. You should reflect ... on the reflection calculations!

It should be noted that one of the principal problems of the above-mentioned topologies, concerning communication, is the inequalities of distance between nodes – which is a great nuisance in relation to TDMA with distributed synchronisation. In fact, because of the different distances between nodes, there are also numerous different propagation delay times between nodes, which in principle could cause the time slots which are reserved for access to the medium by the various nodes to overlap in time.

To overcome that, what comes to mind immediately is a topology in which all the nodes are (on paper) arranged at equal distances from each other, the so-called 'star' topology.

9.5.4   A Star is Born! ... Linkage by Star

9.5.4.1   Passive Star

To avoid the problems mentioned in the paragraphs above (different distances, dissimilar propagation times) and those concerning stubs, their lengths and their relative positions on the network, we tiptoe gently towards a topology in the form of a star, called a 'passive star'; that is, a star-like structure in which it will be considered that the common central point is a 'big welding spot', and the branches of which will be strictly of identical length (see Figure 9.12).

Again, we can always dream! In this case, all the nodes will play practically the same role – at close to the distance at which they actually are from this common point. In principle, if all the nodes were at the same distance from the central point, the propagation times would be identical from node to node, but taking account of the fact that, for numerous practical and mechanical reasons, that never happens in real applications, it is also necessary to specify certain values to be observed, to make this type of topology actually functional:

Maximum distance between two nodes in a system	24 m
Maximum number of nodes in a possible star system	22

It should be noted in the figure that only the two nodes at the most distant ends of the network have line terminations to close the impedances.

Obviously, in all the topologies described above (linear buses, passive stars), in the case of mechanical disturbances on the wires (short circuits between wires, short circuits of a communication wire relative to earth, and so on), the whole network is faulty and communication is no longer ensured at all (see Figure 9.13).

Passive Star

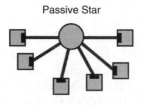

Figure 9.12

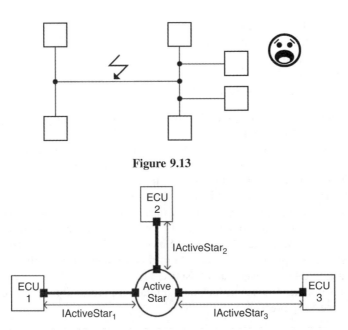

Figure 9.13

Figure 9.14

To avoid that, it is usual to think about insulating or securing the whole or parts of the network from the other nodes, by designing and establishing separations or decoupling of part of the network, using topology which is built around stars which are no longer passive but now active. That is what we will examine now.

9.5.4.2 Active Star

In the paragraphs above, the star which we described was passive, since it was totally incapable of interpreting anything. In the present case (see Figure 9.14), the star includes, or can include, on-board 'intelligent' electronics, which can be used for multiple functions. Examples are routing a message to the right node(s), or disconnecting one or more of these branches in the case of malfunction of a link, and so on. Of course it can also give some muscle tone (the repeater function) to a signal which was a little out of breath. Because this star is now electronically active, it must also be considered as a true line termination, and therefore include a line adapter on each of its ports.

This practice also has its limitations in the context of the applications under consideration. Some values to be observed are also specified:

Length of a branch of a node to an active star	24 m
Number of branches of an active star	Minimum 2
The maximum value is undefined	

... which still leaves room for a lot of imagination and numerous possible topological architectures of the network and improvements for components!

It should be noted that the use of an active star does not imply structurally any concept of system redundancy, but only additional possibilities for a guarantee or reliability of operation of certain nodes on the network.

9.5.4.3 Cascade(s) of Active Stars

We are beginning to go into the galactic topological complexity of FlexRay, but it is true that use of this case is common.

In fact, in a motor vehicle, for numerous mechanical or functional reasons and so on, it is never easy to arrange the CPUs mechanically where one wishes, the components to be controlled and driven being often arranged in quite fixed places (... the wheels are not usually inside the passenger compartment, or the windscreen wipers in the boot ... ☺). So to avoid 'stubs', to have clean signals, known propagation times, and so on, the network architectures are often built around solutions based on cascades, not to say rainstorms, of active stars. In short, a true Milky Way! Since a good drawing is worth many words, Figure 9.15 shows a representative example of this topology. Each entity of the network is linked by a single bus between active stars.

As usual, this practice too has its technical limitations. Again, be careful! Certain values are also specified very precisely:

Maximum number of active stars on the route of the signal between any node and any other node of the network	2
(Electrical) distance between two active stars	24 m

As you have just noticed, despite our great enthusiasm, the maximum number of stars is two! This is principally due to the return time of the stars (see the 'truncation' phenomenon and the TSS parameter) and to the asymmetrical propagation delay times, which have already been mentioned several times. Despite that, this opens up new horizons for numerous possible network architectures, and the manufacturers of integrated circuits of active stars are doing everything to free themselves from these parameters so that several (3, 4, ...) active stars can be cascaded.

9.5.4.4 Hybrid Solution

The cherry on the cake! The recipe for the cake? Take all the solutions presented above, stir skilfully over a low heat for a few minutes, and that gives, for instance, the solution

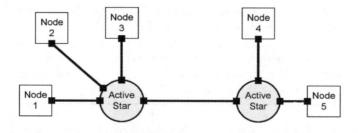

Figure 9.15

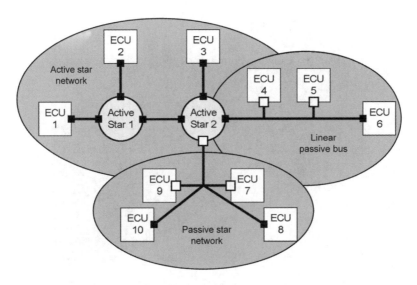

Figure 9.16

shown in Figure 9.16, of course complying fully with the constraints described throughout the sections above.

9.6 Examples of Topologies

To complete this second part of this chapter and give some application ideas to readers, the following paragraphs present some informal examples of network topologies, with some comments. These are conventional examples of mixtures of the topological solutions which are stated in this chapter, and which are used in the same system (in some cases with a single channel, in others with dual channels), and in which sometimes redundancy (total or partial) has been considered.

Since each user and application has its constraints (mechanical, costs, and so on), it is up to everyone to reflect and hope to find (let's be positive and optimistic ...) the solution which is best matched to what is desired.

To illustrate what we say, for the presentation of the examples below we have taken as a common theme the implementation of a Brake-by-Wire topology – and being very serious for once, any resemblance to any implementation whatever would be purely coincidental, the explanations below being for teaching purposes only.

9.6.1 Example of Application for a 'Brake-by-Wire' Solution

Electromechanical braking (EMB) systems, also known as 'Brake-by-Wire', replace conventional actuators with units controlled by electric motors, to link the brakes of the four wheels to the brake pedal, and to communicate between them (see Figure 9.17).

Such systems eliminate the use of vacuum boosters, master cylinders, and so on, and provide better checks on the hardness and stiffness of the pedal, traction control, the

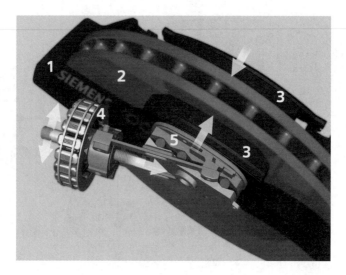

Figure 9.17   Reprinted by permission of Siemens-Continental

stability of the vehicle, the distribution of the braking force, and so on than those found in conventional hydraulic systems. Also, in addition to the above-mentioned benefits, those related to economical and environmental aspects should be added, for example:

- the hardware and software development tools (for more detail see Chapter 19) are designed to reduce the 'time to market' and the development costs;
- the implementation of a FlexRay system in the 'host' vehicle (the production model) is much simpler and much faster than a conventional hydraulic system;
- the environmental problems (brake fluid, recycling of materials, and so on) associated with traditional hydraulic brake systems are eliminated.

Obviously, a conventional hydraulic brake system has a mechanical or hydraulic backup device. An EMB system does not! Consequently, the reliability of an EMB system is absolutely critical, and the system must use a 'fault-tolerant' communication protocol such as that offered by FlexRay (dual communication channel, fault management, bus guardian, and so on).

9.6.1.1   Example 1 (see Figure 9.18)

This example shows a topological configuration using two transmission channels, A in grey, B in black, making it possible to design a system with redundancy.
    For example:

- in the case that channel A, which carries all the braking and ABS (or anti braking system) information, is damaged at any point (for example following a short circuit), it becomes totally inoperative/failed, and in this case channel B can take over and carry the same information – and vice versa for channel B;

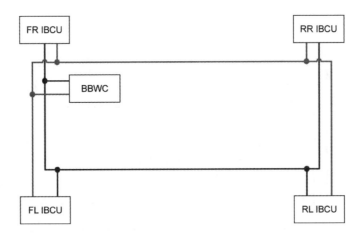

Figure 9.18

- in the case that channel A carries only braking information and B carries only the necessary additional information for the operation of the ABS part of the brake system, if A fails the braking information of the channel can be switched to channel B, thus defining a fallback position of braking without ABS.

Another topological point, which seems naive but is not as innocent as it seems. Taking channel A as an example, let's ask the following question:

- Are we in the presence of a topology of bus type, with three small stubs?
- Or are we in the presence of a topology of bus type, with two small stubs and one large stub?

The answer to this question, which seems remarkably stupid, is entirely hidden in the position of the places where the implementations of the resistors for matching line impedances of the bus physically are (. . . as usual, not shown in the figure, otherwise the joke is lost!), so:

- if the resistors are placed at bottom left and top left, you are in topological situation a);
- if the resistors are placed at bottom left and top right, you are in b).

. . . with all the worries about two different topologies!

9.6.1.2 Example 2 (see Figure 9.19)

This example, which, by its structure, is fully capable of managing redundancy since all the nodes are connected to each other by two distinct communication channels, is built around two active stars, so that if one branch of one of the stars fails, it is possible to keep all the others active. Obviously, the cost associated with this example is higher than that of Example 1, but may very well be suitable for a structure associated with a top-range vehicle.

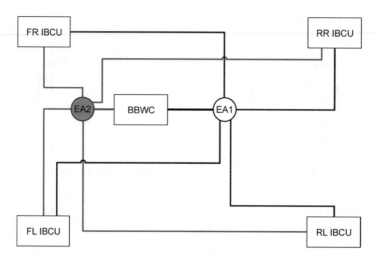

Figure 9.19

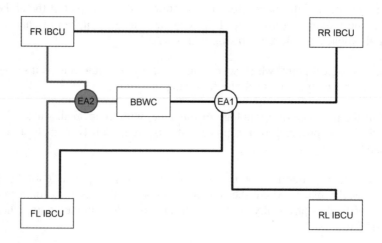

Figure 9.20

9.6.1.3   Example 3 (see Figure 9.20)

As you can see in the figure, this third example is also built around two active stars, but provides a fully redundant structure only on the front axle of the vehicle, which in general supports the greater part of the braking.

Obviously, the cost associated with this example is less than that of Example 2 – the second channel being judged to be unnecessary for the rear axle – but may very well be suitable for an economical structure associated with a mid-range or bottom-range vehicle.

Now that these first examples have been presented, it is now up to you to build your own architectures.

We have now summarised in a few lines the latent topological complexity associated with networks operating at high speeds, FlexRay being one of the worthy representatives for industrial, in particular automotive, applications.

Let us now go on to the part which concentrates on the reception and processing of the signal.

10

Reception of the FlexRay Signal

Land ahoy! The signal finally arrives in sight of a node!

10.1 Signal Reception Stage

At the receiving part 'Rx', the purpose of which is to receive the signal, obviously the same elements are found as mentioned for the transmitter 'Tx', but of course arranged in the inverse order of appearance, that is the ESD device, the EMC filtering and line impedance matching. The same comments as mentioned previously are, of course, certain to apply, and we refer you to Chapter 8 for details of these elements and their application consequences.

Unfortunately, every day, numerous other reasons for electrical degradation of the signal cause a failure to comply (an asymmetry) of the relative positions in time of the leading and trailing edges of the received signals. In fact, because of the properties of the elements which form and participate in the implementation of the physical layer (the medium, the line driver stages, the active stars, the cabling wires, the topologies which are used, and so on), it is often the case that the values of the rise times of the propagated signals differ from those of the fall times.

Let's look quickly at the principal causes for this type of phenomenon, and let's now describe some important details concerning, specifically, the internal structures and performance of the reception stages.

10.1.1 Triggering Threshold

In general, for numerous reasons (optimisation of the signal-to-noise ratio, and so on), to decide or not on the effective presence of a bit, the reception 'Rx' input stages of the incoming signals are usually equipped with detection devices with a triggering threshold and a low but existing hysteresis. The electrical signal which follows this detection and represents the received bit is unreliable and worse, causing different delays between the appearance of the leading and trailing edges, and consequently causing 'asymmetrical propagation delay' distortion, and consequently in turn causing the duty cycles of the received signals to be non-compliant and unacceptable relative to the transmitted

FlexRay and its Applications: Real Time Multiplexed Network, First Edition. Dominique Paret.
© 2012 John Wiley & Sons, Ltd. Published 2012 by John Wiley & Sons, Ltd.

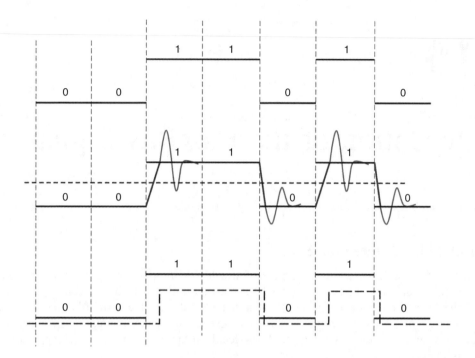

Figure 10.1

signals. In short, the transmitted signal lacks conformity. Figure 10.1 shows this situation, which is very frequent and in fact quite ordinary, and shows the most realistic phenomena of the propagation of a signal on the network and their consequences, called 'asymmetrical effects'.

10.1.1.1 Asymmetrical Effects

Let's quickly list the principal reasons which can cause these asymmetries of the rising and falling signals:

- the presence of an asymmetrical capacitive coupling between the tracks of the printed circuits relative to the propagation lines of the network;
- the asymmetry of the capacitance values of simple wires, linking cables, their positions in the connectors, and so on;
- mismatches of impedances of the terminating loads of the network, or rather the fact that they are not shared symmetrically. These two points cause the appearance of reflected waves, coefficients of reflection, standing wave ratios and therefore asymmetrical distortions of the propagated signal;
- similarly the reductions of slopes by signal reflections, the presence of ringing phenomena (overshoots or undershoots following signal transitions) and crosstalk in the strands of cables produce asymmetries in the signals;
- the electrical imbalance of the components intended for protection from ESDs, which are arranged on the lines or at the line termination;

- too high values of the leakage/stray inductances and/or hysteresis phenomena of the inductors (called 'common mode chokes', which are arranged on the lines nearest the outputs of the line drivers), which are intended to reduce the EMC. Their symmetries and complementarities should be better than ±0.25 ns;
- the often congenital asymmetry of the physical implementation of the output stages (impedances of outputs different between the high states and low states of push–pull stages) of the line drivers and/or active stars can cause asymmetries and different values of the rising and falling signals. Additionally, these asymmetries can vary and evolve with variations of temperature and power supply voltage – but can be considered to be constant during reception of a frame;
- and so on.

In the case of FlexRay at 10 Mbit/s, a sequence of logical bits 1 0 1 0 1 0 1 0 1 0 1 ... with NRZ bit encoding (... therefore giving a 5 Mbit/s wave of square signals), the duty cycle of the received signals should be '50/50' (100 ns ON, 100 ns OFF). Unfortunately, according to the degrees of alteration and asymmetrical deformation of the leading and trailing edges of the signal and the values of rise time, fall time and other heights of thresholds, hysteresis, and so on, it is, for example most often 60/40 ($100 + 20 = 120$ ns ON, $100 - 20 = 80$ ns OFF) or 40/60 ($100 - 20 = 80$ ns ON, $100 + 20 = 120$ ns OFF). An example of this situation is shown in Figure 10.2.

As we shall see in Chapter 11, the presence of these asymmetrical faults causes serious violations of the eye diagram to which it is necessary to conform in order to guarantee that the network functions well. For this purpose, the FlexRay specification indicates that the asymmetries are limited to or tolerable up to 4 ns for the transmitter, 5 ns for the receiver and 8 ns for an active star.

10.1.2 Unique Effects at the Start of Transmission and/or Reception of Frames

Obviously, daily life is not always so simple!

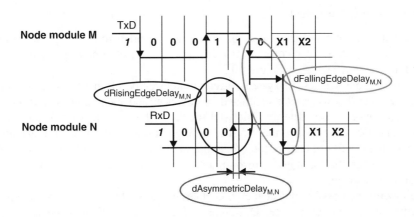

Figure 10.2

Often, before transmitting or receiving – that is before putting itself into a state for functioning correctly – a system has a certain inertia to the initialisation or reception of the propagation of the signal. A particular effect then occurs, only at the start of a new signal transmission sequence, and not at all in the course of it. This is the case in many communication networks, and in particular in those that operate under FlexRay. Let's look quickly at two very classic scenarios.

10.1.2.1 Starting a Communication

At the time of initialisation and validation of the start of a communication in its time slot (in this case, the start of a FlexRay frame), the node controller under consideration, going from a position of listening to the network during the preceding slots to a position of message transmitter (or the inverse during reception), begins to put its line control stage (the 'bus driver') into operation. This takes a certain time to put itself into active mode, and consequently can offset, or above all truncate, all or part of the appearance of the first edge of the communication on the network, but obviously will have no effect subsequently on all the other edges to come.

10.1.2.2 Return Time

Being very secretive, so as not to confuse matters we didn't point this out a few paragraphs back, but strictly speaking it is the same when, for example, the signal passes through 'bidirectional repeaters' or 'active stars', since these take a little time for reflection, called 'return time', to determine the direction of the signals passing through them when data are exchanged between nodes of the network.

Having to include reliability elements such as 'active stars' in the network topology if necessary (... necessary for many other functional reasons), and having to take a little time to manage (with or without a microcontroller) the passage of the signal through them, on the one hand adds 'immaterial distances' to the propagation path, and on the other hand possible truncations of the signal.

10.1.2.3 Truncation Effect

As a function of the topological options which are intended for the application and the obstacles (active stars, repeaters, and so on) which it meets on its path, the signal is delayed and/or truncated (see Figure 10.3).

Of course, if one wishes the system to function correctly, all these effects, the times of their occurrence, known and/or estimated in advance, and their finely determined maximum totals (worst case, deterministic, statistical, stochastic, and so on) must be taken into account. Obviously, once the physical configurations of the networks are installed, one hopes that it will not change too often, although that sometimes happens in industrial applications when the topology is reconfigured. It is quite rare in automotive or avionic applications, however.

To allow for the possible physical effects of 'truncation', the FlexRay protocol provides a specific device and procedure. Designing clever devices which make it possible to

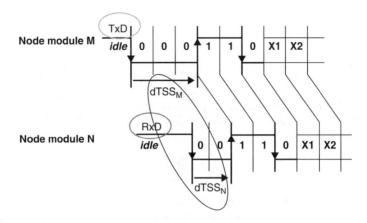

Figure 10.3

reduce or eliminate (artificially) all or part of the consequences of these effects, which are produced only when frames are started due to the participants which are or can be arranged on the network, is then conceivable. In short, the principal trick consists of adding to everyone a large initial layer of modifiable delays, so that a little can be withdrawn sparingly where necessary, so that everyone believes that they all continue to arrive at the same time! (For the details, look back at the specific sections concerning the TSS parameter and the 'action points' in the description of the encoding of the FlexRay frame).

NOTE

As indicated in Figure 10.3, the phenomenon due to the propagation delay of the signal and the 'truncation' effect which can occur do not have a cumulative effect on the timing.

10.1.3   Summary of the Effects of Truncation of the Complete Chain from Tx to Rx

Let's now quickly make a list of the various elements and phenomena which can cause deterioration of the integrity of the signal.

10.1.3.1   Asymmetrical Effects

At any time, numerous reasons for degradation of the electrical signal can cause non-compliance (asymmetry) of the position of the leading and trailing edges of the received signals in time, and consequently final modification to a greater or lesser extent of the duty cycle of the received signal which must be processed by the receiving CC, and which sooner or later can cause higher BER values.

Figures 10.4 and 10.5 show the effects and maximum authorised 'asymmetrical' participations for each of the participants throughout the length of the communication chain,

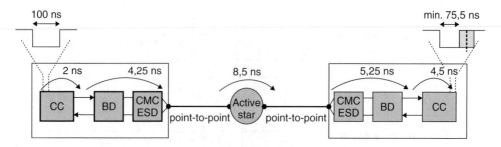

Figure 10.4

from the part that outputs the message to the medium in the strict sense, the topology and thus various elements which are present on it, to the reception part, its small effect on the possible asymmetrical degradation of the signal being at stake for each of them, as it should be!

10.1.3.2 Asymmetrical Delays ('Worst Case')

CC (sender)	2 ns
BD (bus driver) + ESD + coil	4.25
Active star	8.5
Coil + ESD + BD	5.25
CC (receiver)	4.5
that is a total of 24.5 ns ➜ 'bit min' = 75.5 ns	

The FlexRay specification indicates that a 'bit min' value of 62 ns is tolerable – leaving a safety range of 13 ns for repercussions of the stochastic effects due to EMC problems on the asymmetrical delay.

One last point: as we indicated above, the propagation speed on the medium is often better (of the order of 6 ns/m) than what is expected in the official specifications (10 ns/m).

10.2 Processing of the Received Signal by the Communication Controller

When the signal is received in a node, after passing through the Rx part of the line driver, the received bits and frames are decoded, validated and interpreted at several levels in the CC (see Figure 10.6). Additionally, the CC must carry out various tasks, in particular that of cleaning the incoming signal when disturbance or noise of any kind is present, and resynchronising them. Let's look in detail at the first two principal processing stages, which are acquisition and bit adaptation.

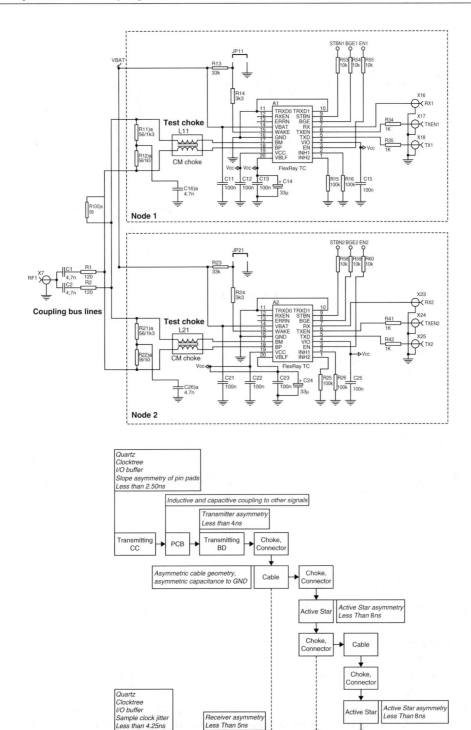

Figure 10.5

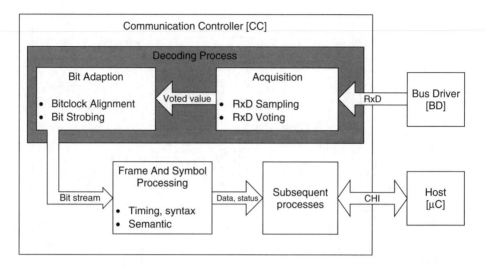

Figure 10.6

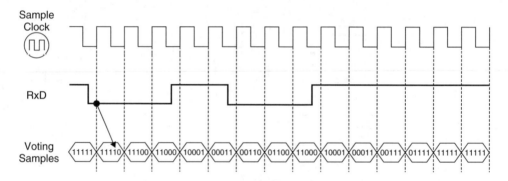

Figure 10.7

10.2.1 Acquisition of the Binary Flow

First, the flow of bits from the Rx output of the line driver is:

- acquired by the CC;
- packaged and reformatted;
- sampled at each rising edge of the local 'sample clock' at 8 samples per bit (*cSamplesPerBit*: =8);
- on the same occasion, the noise (glitches) is also acquired in the same way;
- the samples thus obtained are recorded and transferred into a buffer memory, called the 'voting window'.

10.2.2 Suppression of Disturbance/Noise

To suppress (deglitch) noise during the bit, in accordance with the FlexRay specification, the CC proceeds by a weighted vote technique with majority logic, called 'RxD voting'. To satisfy that (see Figure 10.7), after sampling:

- the samples are directed to a reception buffer memory which is organised as a first in, first out (FIFO), of depth 5;
- all new incoming values thus remain for five clock ticks in the 'voting window' of the FIFO;
- by the principle of operation of a FIFO memory, this organisation produces a voting window which floats in time;
- the ratio of measured values is formed over the five recorded values;
- the voted value ('1' or '0') of the output signal is the value which is in the majority among the five recorded values (so at least three out of five); consequently, disturbances which last for less than three samples – that is a quarter of the duration of a bit (at 10 Mbit/s that corresponds to 25 ns) – can be suppressed;
- the voted value of the output signal is then always determined at the rhythm of the local clock, and is not yet synchronised;
- by structure, this method introduces an interpretation delay of a value of two samples (*CVotingDelay* $= 2$).

10.2.3 Binary Alignment

The local bit clock is then adjusted and aligned (the procedure called 'bit alignment') to the flow of data (the voted values above), and then the official value for logical processing of the bit is determined (bit strobing).

10.2.3.1 Bit Clock Alignment and Bit Strobing

Now, in order to synchronise the receiver node with the transmitter node, the falling edges of the data flow of voted values are used as a reference point, and the sample counter is reset and initialised, not to 0 but to 2. This makes it possible to synchronise the internal local bit timing to the flow of incoming data, with the resolution/granularity of the local node's own sample clock.

From this instant, the bits are considered to be strobed, and take the official values of the logical data carried by the frames of the protocol, Data_0 (low bit) and Data_1 (high bit).

This last data flow (Data_0, Data_1), which is now synchronised, can then be taken in hand for subsequent processing, which consists of verifying the timings of formats, the syntax, the semantics of the frames and other transmitted symbols. This is what is called 'frame and symbol processing (FSP)', and the CC, with help of the host processor of the CPU, then proceeds to decode the binary content of the received frame, in order to manage its application.

11

The Bit Error Rate (BER)

Here we are, almost at the end of our troubles.

Now that we know all the vicissitudes of the integrity of the signal and the electrical and electronic mysteries of the transmitters (Tx), medium, topologies, input stages (Rx) and bit decoding of the network – at least by name – there are only two *small* details to resolve! The first, that of displaying the nominal bit error rate (BER) (about which no-one could care less . . .), giving the nominal qualities and performance of the communication network, and obviously, above all, the second, corresponding to the value BER_worst_ case, with all dispersions, tolerances, and so on mixed up – about which everyone cares a great deal!

So to work, for the final ordeal of this third part!

11.1   Integrity of Signal and BER

Now that we hold all the mysteries of the topologies and the consequences of their drifts and variations, we are able to describe all the phenomena that cause non-integrity of the received signal. Two well-known diagnostic tools are commonly used to analyse and quantify the integrity of the signal and its consequences. These are principally the 'eye diagram' and the 'bathtub curve' associated with it.

11.2   Eye Diagram

The eye diagram and the quality of its opening are the conventional elements which make it possible to indicate the performance of the signal in communication techniques and digital transmission. Use and interpretation of it make it possible to demonstrate and estimate the performance regarding the integrity of the signal, and then to calculate and/or evaluate what the BER in relation to communication could be. It is part of the elements which summarise the overall quality of communication. It is therefore unsurprising that the FlexRay specification indicates that ' . . . *as long as the differential output voltage satisfies*

FlexRay and its Applications: Real Time Multiplexed Network, First Edition. Dominique Paret.
© 2012 John Wiley & Sons, Ltd. Published 2012 by John Wiley & Sons, Ltd.

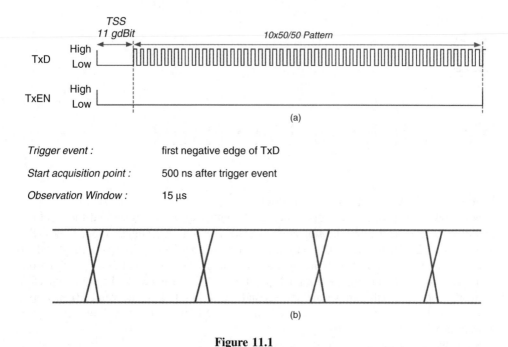

Figure 11.1

the eye diagram given in the FlexRay specification, the effect of the network topology is of no importance ...'. That at least has the merit of being clear.

11.2.1 Brief Reminder

The eye diagrams are obtained using actions which are carried out using a sampling oscilloscope with a persistent screen, and consist of displaying simultaneously on its screen the superposition of multiple traces of data bits, triggered by a bit clock. In this case, thanks to its persistence, the screen indicates the envelope of the amplitudes and timing fluctuations of the signals, and the central region of the figure which is obtained resembles an eye – which is why it is called an eye diagram.

Figure 11.1a shows the signals to be applied (the stimulus), which are in fact a sequence of square logical digital signals 1 0 1 0 1 0 1 0 1 ... coded in NRZ, and Figure 11.1b shows the obtained eye diagram (a wave close to an ideal, undistorted square wave, since there is no filtering, and with rise times and fall times of finite value). As was hoped, the opening of the eye is broad and high.

11.2.2 Jitter

Reality is quite different, since the signal always undergoes some deterioration of timing and/or amplitude:

- On the one hand, the rising and falling edges are always more or less subject to so-called 'jitter' errors (see Figure 11.2), which occur:
  - either following misalignment of the instants and/or values of the rise and fall times (for example due to noise, crosstalk, clock jitter or a fault of (re) synchronisation);
  - or when the communication speed is fast, the above (absolute) timing errors become predominant and have the effect of closing the eye of the diagram, causing a higher potential risk of errors in the digital data;
- On the other hand, the amplitudes fluctuate as a function of variations of power supply, earth noise, and so on.

As an example, Figure 11.3 shows an eye diagram of the received signals, and is much closer to reality.

This figure shows only qualitatively the range of amplitudes and timing deviations associated with the physical representation of the data. To conclude, an eye diagram which has a large overall opening of the eye indicates that the flow of data is very little

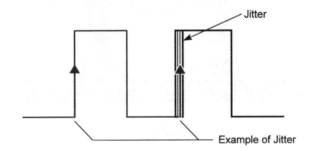

Figure 11.2

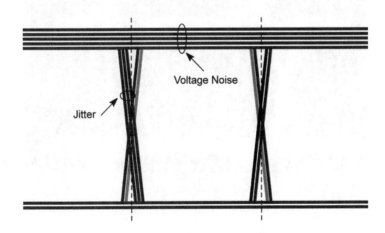

Figure 11.3

subject to amplitude variations, and how little the noise affects its timing. In contrast, a small overall opening of it indicates that the data are very noisy (see Figure 11.4).

Example

While keeping the same assumptions as before (that is, a continuous sequence of bits 1 0 1 0 1 0 1 0 1 0 coded in NRZ), Figure 11.5 shows an (alas typical) eye diagram in the case of the presence of reflections due to imperfect (or bad) matching of the impedance of the communication line.

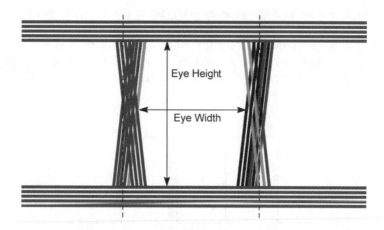

Figure 11.4

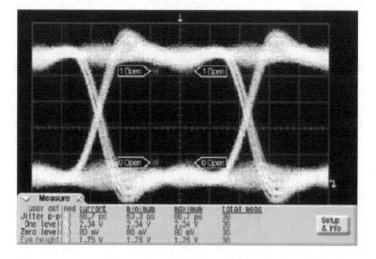

Figure 11.5

If the reflected waves have significant amplitude, they can take part in reducing, in particular, the width of the eye opening, and thus greatly increasing the potential risk of bit errors.

In addition to that, it must also be pointed out that there are very often numerous other, additional errors which are not presented in the preceding paragraphs, including amplitude distortions due to losses in the transmission system, and problems such as crosstalk with other lines which also carry signals (for example those of CAN). This crosstalk (for example reinjection of interfering signals by coupling between linking wires in a strand/harness of wires) can be elusive, since it can imply signals other than those that are wanted. It should be noted in passing that the effects of unsynchronised crosstalk may not be clearly visible when the diagram is displayed.

For information, for the whole communication chain, the FlexRay specification indicates certain limits which should not be exceeded when the impedances of terminating loads equal the mean value of 90 Ω (see Table 11.1). Also, Figure 11.6 indicates the points at which measurements must be carried out.

In the case that the slopes of the signals have values less than the slopes indicated above (because of imperfect matching of receiver thresholds), an asymmetrical effect of up to 4 ns can be accepted.

Table 11.1   FlexRay eye diagram

	At the transmitter	At the receiver
Test point	Measured at TP1	Measured at TP4
Specified voltage	|uBus,min| = 600 mV	|uBus,min| = 400 mV
	|uBus,max| = 2000 mV	|uBus,max| = 2000 mV
Comment	Represents the 'minimum eye aperture' of the transmitter	Represents the 'minimum eye aperture' of the receiver

Minimum aperture uBus @ TP1

600mV	600mV
12.5ns	83.5 ns
0mV	0mV
0ns	96ns
−600mV	−600mV
12.5ns	83.5ns

Minimum aperture uBus @ TP4

400mV	400mV
15ns	65ns
0mV	0mV
0ns	80ns
−400mV	−400mV
15ns	65ns

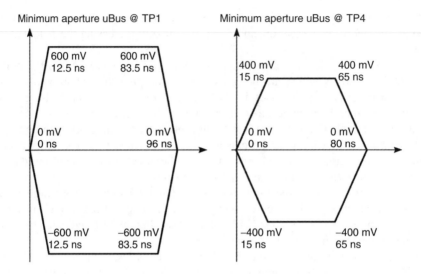

Figure 11.6

11.3 Relationship between the Integrity of the Signal, the Eye Diagram and the BER

Being now in possession of all the measured timing values, it becomes easy, in a few moments (in fact several seconds, using a device of bit analyser type, which works well but is expensive), to construct the histograms (to the left and right of the eye opening) of the measurements which are carried out, and which indicate more clearly the most probable instants at which data transitions occur (see Figure 11.7). Incidentally, this histogram is known as a 'jitter histogram' – of course!

Now, if the scale of the ordinates of this histogram is modified so that the total value of the integral of this curve equals one, it will then become and represent the probability density function (PDF) of the jitter of the observed signal (see Figure 11.8).

Now, in a few lines, let's finish off the maths, which are very important for certain crucial choices about FlexRay.

11.3.1 BER

Ideally, to obtain the best (lowest) BER which is inherent in a given eye diagram, it is preferable that the receiver samples the signal at its centre, the place where, in principle, the tails to right and left of the histograms are smallest, as shown in Figure 11.9.

To calculate the overall probability due to one or the other (to the left and right) of the position errors of the transitions due to jitter of the rising and falling signals causing errors in the interpretation of the bit values, it is necessary to calculate the value of the area below the tail of its PDF function on the wrong side of the sampling point (in time). This integral/integration forms what is called the complementary cumulative distribution function (CDF). For the left-hand part of the PDF, the calculation is carried out so that the

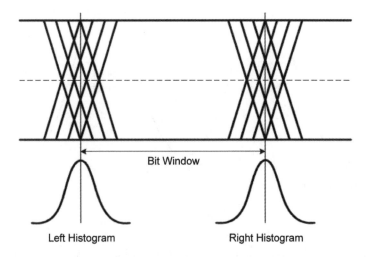

Figure 11.7

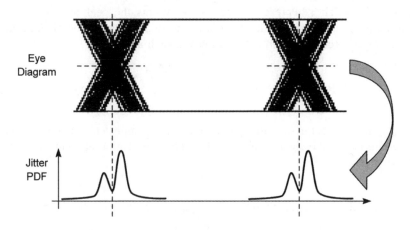

Figure 11.8

integration of the distribution tail is carried out from the sampling point to plus infinity, and for the right-hand part, from minus infinity to the sampling point. The total probability of error due to transitions is, of course, the sum of the two CDFs. It is also accepted that the tails of the adjacent bits do not contribute to the value of the probability of error. This is all the more true for FlexRay bit encoding, which is of NRZ type.

11.3.2 Calculating the BER

To determine the value of the BER, the value of the probability of error due to transitions must be multiplied by the value of the probability of the occurrence of the transitions. Nominally, the latter can be seen as the mean transition density. This model assumes that

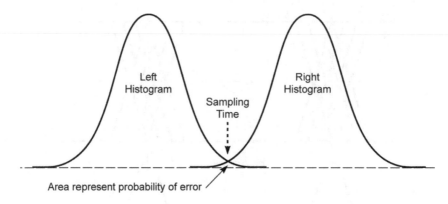

Left
Histogram

Right
Histogram

Sampling
Time

Area represent probability of error

Figure 11.9

typical data flows have a transition density of 50% (a long sequence of 1 0 1 0 1 0 1 0
1 0 ... coded in NRZ).

To demonstrate this concept, usually a generic jitter PDF JT (τ, W, σ) is defined,
centred at 0, in which τ represents time, W represents the peak-to-peak value of the
deterministic jitter, and σ represents the root mean square value of the random jitter. The
left PDF histogram (centred at 0) results in errors such as

$$\text{BER}_{\text{left}}\left(\tau_{\text{sample}}, W, \sigma\right) = \Gamma_{\text{transition}} \int\limits_{\infty}^{\tau_{\text{sample}}} JT(\tau, W, \sigma)\mathrm{d}\tau$$

where τ_{sample} represents the sampling point, and $\Gamma_{\text{transition}}$ represents the transition density.
The same is done on the right (shifted to the right and centred at 1 unit interval), with
the following result:

$$\text{BER}_{\text{right}}\left(\tau_{\text{sample}}, W, \sigma\right) = \Gamma_{\text{transition}} \int\limits_{\infty}^{\tau_{\text{sample}}} TJ(\tau - UI, W, \sigma)\mathrm{d}\tau$$

These results are summed, finally giving:

$$\text{BER}_{\text{total}}\left(\tau_{\text{sample}}, W, \sigma\right) = \text{BER}_{\text{left}}\left(\tau_{\text{sample}}, W, \sigma\right) + \text{BER}_{\text{right}}\left(\tau_{\text{sample}}, W, \sigma\right)$$

First, it must be understood that the jitter PDF represents the integral corresponding to
the product of convolution between, on the one hand, the deterministic jitter function,
which itself is bounded, and on the other hand the Gaussian random jitter function, which
itself is unbounded. Because the deterministic jitter function is finite and bounded, only
the extremities of the jitter PDF are formed by the distribution 'tails' of the Gaussian
function. The effective standard deviation (the classic root mean square value) of the
Gaussian function can then be calculated/extrapolated using the conventional method
of least squares, which is very suitable for the regions of these distribution tails. This
extrapolation makes it possible to determine the value of the probability density of the
very rare instants of data transitions which cannot be captured by the system because

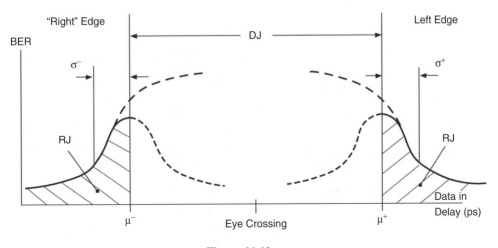

Figure 11.10

of its own time measurement limitations, and which consequently can cause bit errors (see Figure 11.10).

11.3.3 Bathtub Curve and BER

Once the jitter PDF is determined from the histogram of measured jitters and the above-mentioned method of least squares has been applied, the corresponding curve can be estimated – and it's in the form of a bathtub! (That was just to put you back in the bath in case you weren't following!) As a reminder, the bathtub curve represents a dimensionless number which indicates the value of the probability of bit errors (the BER). The indefinite integral of the full PDF jitter curve represents the bit errors due to timing variations. Therefore, the bathtub curve of timing errors is simply the integral or CDF of the PDF jitter curve.

Consequently, the probability of closure of the eye – for the left-hand part, the PDF jitter function is integrated from right to left.

$$\mathrm{BER}_{\mathrm{right}}(t) = \int\limits_{-\infty}^{t} \mathrm{PDF}(t')\mathrm{d}t'$$

and similarly, for the right-hand part, the PDF jitter function, the integral is calculated from left to right.

$$\mathrm{BER}_{\mathrm{left}}(t) = \int\limits_{t}^{\infty} \mathrm{PDF}(t')\mathrm{d}t'$$

Figure 11.11 summarises the last few paragraphs, while relating the variations of PDF and those of BER.

In the glimmer of the above explanations and of many others which can be found on the Net, we hope that each of you will have understood that it is your responsibility

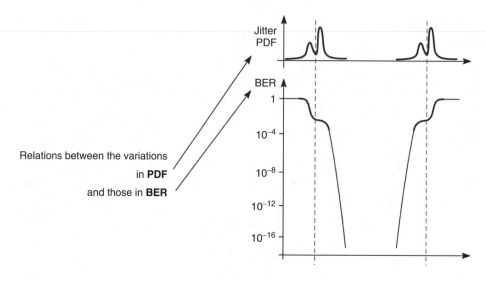

Figure 11.11

to define the BER which you want for your application, and therefore to be able to go back to the PDF curve, and to its associated and desired eye diagram, and finally to the acceptable jitter for your system.

To conclude these few lines of reminder, the eye diagram is a simple and very instructive tool, which makes it possible to evaluate finely the circuits and systems of digital transmission. Used as a complement to other tools for measuring the integrity of the signal, it can help to predict future performance and to identify the sources of system errors. For information, a good FlexRay development cannot do without such a study.

To have a first idea and to evaluate the BER, let's model and simulate the network and all the elements which compose it.

12

Modelling and Simulating the Performance of a Network

12.1 Modelling and Simulating the Performance of a Network and its Topology

Nowadays the complexity of networks does nothing but increase, with the presence of more and more electronic control units (ECUs), variations of topology,[1] and so on. Obviously, all that complexity drastically increases the requirements for verification that communication is functioning well, and consequently, users are confronted with the need to verify, very early in the design of a system, the numerous architectures, the integrity of the signals, the tolerance to faults as a function of the different loads in different fields and environments, and so on.

Verification means having a methodology for validation, and obviously there is a choice of them. The first that comes to mind consists of carrying out series of measurements (in fact batches) in the laboratory, to cover all the possible cases. You can dream, but you will quickly realise that the field of parameters to be made to vary is immense, and to cover them completely by systematic measurements would take far too much time, and the cost of them would explode all the budgets of the planet!

What remains, as the final way out, is to simulate the whole solution, components, topologies, characteristics of the network, and so on while crossing your fingers (and sometimes your toes) that everything goes well.

12.2 Modelling the Elements of the Network

Simulating is all very well, but you still need to have good simulation models! Let's begin by asking where good models are available.

---

[1] Certain applications support topologies which are almost fixed, frozen or stable. This is the case, for instance, of vehicle design 'model by model'. Obviously, designing them is not simple, but at least is well delimited.

Other applications are subject to very fluctuating topologies, with highly (often very highly) variable geometries. This is the case, for example, of the design of vehicle "platforms" on which it is intended to deploy a multitude of derived models, options, etc., and the design takes a long time and is tricky. To mention just one example of a problem: where physically do you arrange the line termination matching resistor when, depending on the options of the model, you don't know where the end of the line will be? And there are dozens of others to solve.

FlexRay and its Applications: Real Time Multiplexed Network, First Edition. Dominique Paret.
© 2012 John Wiley & Sons, Ltd. Published 2012 by John Wiley & Sons, Ltd.

12.2.1   Simulation Models

In general, it is always quite tricky to find suppliers of models which are very representative of their products, either because the demand for this specific type of product has never existed before and no-one has ever spontaneously had the idea of implementing a model, or because by giving a realistic model, the component manufacturers would risk unveiling a few too many technological secrets about the implementation of their products! So there are often models which are 'close' to reality, but which still conceal numerous small shady areas and doubts about the truth of their performance. That being so, we have made almost superhuman efforts to obtain some information, and more or less succeeded.

Now let's begin our list.

12.2.2   The Line Driver/Transceiver

Figure 12.1 shows an example of a simulation model of a line driver which is well known in the business. It is certainly not the most correct – but on a given date, it's the nearest one which is available! These diagrams, which are described in certain languages specific to different simulation tools (P-SPICE, SABER, and so on), are obviously accompanied by all the values of their parameters.

12.2.3   The Communication Line

Who dared to say that the communication line consisted of nothing but ordinary wires mounted as twisted differential pairs, screened or otherwise? An example of a diagram used to model it is shown in Figure 12.2. It should be noted that the values of the series resistors depend on the frequency through the skin effect, which, depending on the rank of the harmonics included in the signal, will not fail to distort the response and performance of the medium.

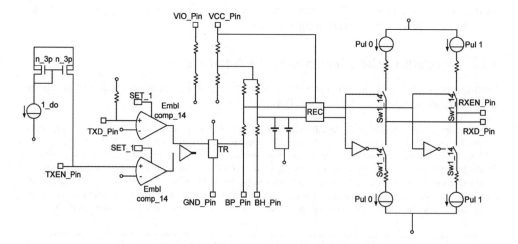

Figure 12.1

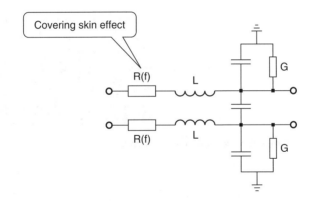

Figure 12.2

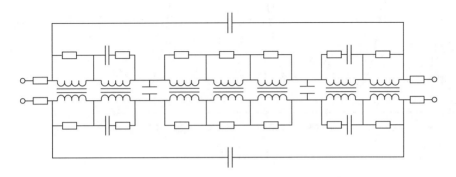

Figure 12.3

12.2.4   The EMC Filtering Dual Inductors

Here, too, what a surprise! (see Figure 12.3). All that for a miserable little filtering dual inductor! It's true, we forgot to say that its winding had to be a little special, for example of bifilar type, so that its symmetry should be as perfect as possible to minimise the asymmetrical propagation delay and comply with its maximum value, and so on. In short, the most perfect agreement of the actual results compared with the simulations is at this price.

After this long stage of modelling the various elements, you're very proud of yourself, and you finally launch the simulation and wait . . .

Just to convince yourself that the simulation model is good and that the values of the parameters are realistic, you carry out a few series of measurements . . . and you don't just cross your fingers again, but all your toes at the same time!

12.2.5   Back to Reality

It's a great moment of truth when you make comparisons between the results of the measurements and those of the simulations! Mr Murphy already said it, '*If it's strictly*

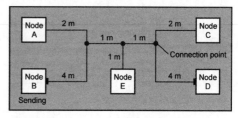

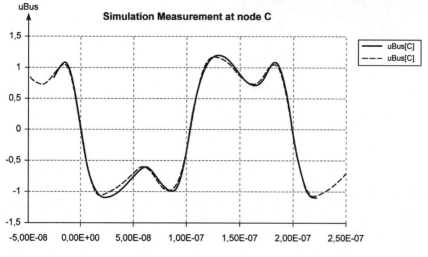

Figure 12.4

equal first time round, it's because there's a double error somewhere!' Obviously, it's never equal. It's never very far, but you always have to refine the modelling. At the end of numerous convergent (if possible) iterations, you generally arrive at something which is almost consistent. (In fact, you have broadened and optimised your knowledge of the model, simply to make the results of the fine theoretical simulation 'stick' to the sombre practical reality!)

After these few reminders and theoretical remarks, and since nature abhors speaking in a vacuum, Figure 12.4 provides (in a continuous line) an actual example of signals observed on a conventional FlexRay network – with, as if by chance, a strong family resemblance to those of Figure 12.5. Weird! It should also be noted that the predictions of the simulation (dashed) are very close to reality, and from that it should be concluded that the model which is used is only a short step away from reality.

Figure 12.6 broadens the field of vision of the problem and gives a view of the signals (which are far from square) in various nodes, depending on the topology of the same network.

Here we are, almost saved!

12.3   Simulation

It's the 'almost' of the last sentence that we could do without! Unfortunately, very unfortunately, to evaluate and predict reliably the performance of the networks and systems, it

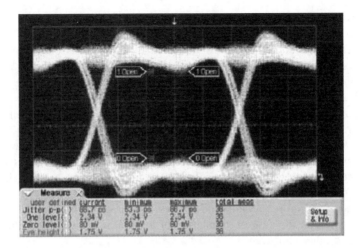

Figure 12.5

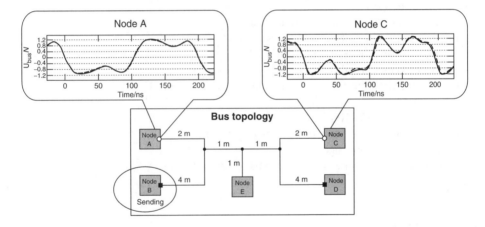

Figure 12.6

is necessary to take account of all the tolerances (min, max) and the limit values of the components. Also, that must be done quickly and at low cost. All you have to do is scan through all the variations of the parameters, but how? On the table, using a PC, using your nose, by brute force, randomly, scientifically? The suspense is unbearable! In any case, for the system to function correctly all the time, you must succeed in finding the conditions where the least favourable 'worst cases' are.

12.3.1 Visit Monte Carlo, its Rock, its Casino and its Method

When you hit a rock, the method called 'Monte Carlo' (guess why), which aims to calculate numerical values using random methods – that is probabilistic techniques – is often used.

In a few words, this method was devised in 1947 by Nicholas C. Metropolis. It has this name because it is based on using random numbers and statistical sets (compare with the games of roulette and cards in casinos). Its principle is based on generating a random sequence of accessible states (for purists, what is called a 'Markov chain') in the space of configurations of the system being studied. The values that one wants to use are sampled, privileging the regions where the probability density of the canonical set in this space is highest. A property of equilibrium of the probability is then obtained, as a simple average over the accepted configurations. This method is widely used because it represents a simple and relatively efficient means of obtaining averages of physical magnitudes in a statistical set. It is important to note that these averages are obtained despite the impossibility of knowing explicitly the normalised probability density of the set under consideration.

This method is therefore suitable and always indicated, to have a more precise idea of the 'worst cases' in the case of simulation of a FlexRay network including electronic components and other hardware.

For example, parameters such as the following are taken as variables:

- for the line driver:
  - the input and output min and max voltages,
  - the threshold min and max voltages,
  - the min and max rise and fall times,
  - the max asymmetrical delay,
  - the input and output min and max capacitances 10–50 pF;
- for the cables:
  - the impedances 80–100 Ω,
  - the capacitances and inductances per unit length;
- for the passive components:
  - the min and max values of the termination impedances;
- and so on.

12.3.1.1   Results

Figure 12.7 shows an example of the cartographic representation of results of a particular variable (for example propagation delay time including the asymmetrical delays of the network), obtained as a function of variations of one or more specific parameters cited in the list above.

It is now up to everyone to draw conclusions from their own simulations.

12.3.2   Examples of Performance and Recommended Topologies

Sparing you many other considerations (costs, and so on) and not claiming to have innate knowledge, Figures 12.8 and 12.9 (from NXP Semiconductors) show **one example**. On the one hand, they show us a summary of the performance and/or recommendations of

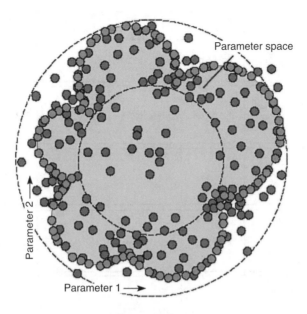

Figure 12.7

	Asymmetric Delay	Propagation Delay	Signal Integrity	Truncation (TXD to RXD)	EMC
Point-to-point (1 — < 24 m — 2)	< 16 ns	max. 244 ns	++	max. 250 ns	++
Daisy chain bus (1 — < 24 m — 2 / 2, 21)	< 18 ns	max. 244 ns	++	max. 250 ns	+
Small linear bus (1 — < 12 m — 5 / 2, 3, 4)	< 20 ns	max. 172 ns	+	max. 250 ns	+
Large linear bus (1 — < 24 m — 22 / 2, 21)	< 20 ns	max. 244 ns	O	max. 250 ns	O
Passive star (1 — < 12 m — < 12 m — 5 / 2, 3, 4 < 1 m)	< 20 ns	max. 244 ns	–	max. 250 ns	–

Figure 12.8   Topologies without active stars

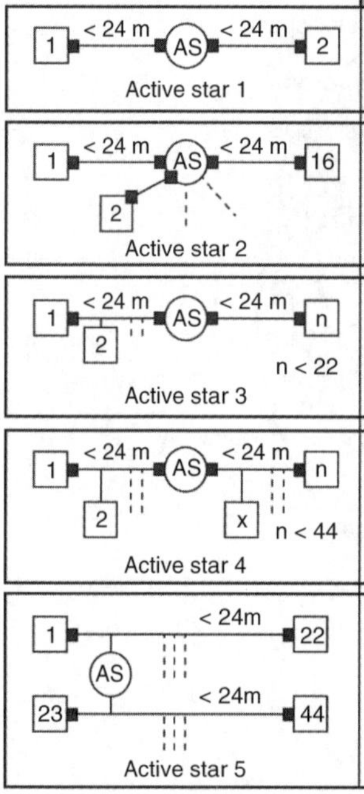

	Asymmetric Delay	Propagation Delay	Signal Integrity	Truncation (TXD to RXD)	EMC
 Active star 1	< 24,5 ns	max. 488 ns	++	max. 700 ns	++
 Active star 2	< 24,5 ns	max. 488 ns	++	max. 700 ns	++
 Active star 3	< 26,5 ns	max. 488 ns	+	max. 700 ns	+
 Active star 4	< 32,5 ns	max. 488 ns	(−)	max. 700 ns	(−)
 Active star 5	< 35,5 ns	max. 488 ns	(−)	max. 700 ns	(−)

Figure 12.9   Topologies with active stars

the principal uses of FlexRay, and on the other hand they indicate the relative effects of different topologies that can be used under FlexRay, relative to the aspects of EMC, propagation delay, asymmetrical delays, integrity of the signal, truncation, and so on. To each his own, and no-one is a prophet in his own country, so we invite you to form your own opinion on this subject.

13

Summary on the Physical Layer of FlexRay

The wide range of possible network configurations and topologies and the ranges of variation of the parameters of its components represent an enormous challenge for the execution and implementation of a robust FlexRay physical layer. Despite that, although it's primary evidence, very often we have the pleasure of repeating to anyone who will listen that there's no point in dreaming about application layers, however well they may perform, and starting to write thousands of lines of code, unless the physical layer functions correctly and is guaranteed. So let's say it again, and again!

On this subject, simulations of Monte Carlo type, or others if they are more suitable, must be seen as indispensable tools for success in building robust network architectures. Also, use of this method makes it possible to go on to even more thorough investigations during the preliminary phases of development of the network on real vehicles.

One last comment about these simulations: sometimes what are called 'corner cases' (special cases) occur in the implementation of networks, and in particular with FlexRay. These are what some people call 'worst cases'. Resolving or trying to resolve them is aiming for perfection, but it is often expensive, and often the 'quality/cost' ratio stops with the first evaluations of the 'risk incurred/cost' ratio.

Finally, it is therefore important, and strongly recommended to future users, to proceed with long modellings and simulations of their parameters (load impedance, and so on) and their topologies before implementing their networks for real – at the risk of painful surprises. It is true that the fundamental questions to be resolved – speeds of 10 Mbit/s, very numerous topological possibilities, management of the worst cases – make it necessary to reflect twice (. . . or three or four times) before finalising the network architecture of a new vehicle or on-board system, and that given the safety aspects at stake, no-one has the right to skip them!

FlexRay and its Applications: Real Time Multiplexed Network, First Edition. Dominique Paret.
© 2012 John Wiley & Sons, Ltd. Published 2012 by John Wiley & Sons, Ltd.

So go to work, to calculate your bit error rates (BERs) as a function of your impedance mismatching, topologies and so forth! It's your turn now!

AUTHOR'S NOTE

Our friends in the profession, designers of all kinds of multiplexed networks, should know that personally, and from now on, we have great sympathy for their present and future suffering!

Part D

Synchronisation and Global Time

We cannot finish this technical presentation, which is dedicated to the FlexRay protocol, without mentioning and describing two thorny problems. The first is that of time synchronisation of the nodes in a so-called 'real time' system for access to the network of TDMA type. The second problem is that of the concept of 'Global Time' (whether in normal operation – Chapter 14 – or during the startup phase of the network – Chapter 15).

The proper thing to do would have been to mention this point during the general presentation of the protocol and the management of frames, but again for teaching reasons, we preferred to defer this important section to the end of the presentation of the protocol. The reason is very simple. Following the previous chapters about the physical layer, you have become aware of the very strong dependence of the propagation times and delays (symmetrical and asymmetrical) and of the deformation of the transmitted signal as a function of the form of the physical layer in terms of numerous variations of possible distances, topologies and media, variable delays because of supplementary filtering which is implemented to reduce radio frequency pollution, in varieties of nodes present on the network, each having on board a microcontroller dedicated to its application, and thus having its own clock, and so on. All that should remind you (for plenty of other reasons) of the long chapters about the problems of synchronisation and resynchronisation of CAN! And in fact, here too, for quite similar reasons, it is absolutely necessary to have synchronisation components, to be certain that the TDMA accesses to the medium – via the time slots of the static segments and the minislots of the dynamic segments – don't tread on each other's toes too much!

Everything that has anything to do with time synchronisation between all the participants of a FlexRay network deserves a chapter to itself. This is, in fact, a key point for ensuring that a real time system functions well, with access to the medium of TDMA type and very variable initial topologies!

Time synchronisation of the nodes/CPUs of a system with distributed intelligence functioning with access to the medium of TDMA type is crucial. As we began to explain in the previous chapters, each node of the network holds within it a specific microcontroller (because of the requirements of its local application), which operates with the aid of the most suitable clock (at x MHz) for its task (with CPUs of types such as PIC, ARM, MIPS, H8, and so on), and the operational sequencing of which depends on its own local clock. Also, the project manager or architect (designer of the architecture of the whole) for the development of the network, for his or her own reasons, chooses a specific communication bit rate and defines the message handling (as a number of time slots and their duration) as a function of the overall application requirements. The game thus consists of reconciling all these timing requirements, which are sometimes finally contradictory. This is the aim of the FlexRay synchronisation mechanism.

So, to work!

14

Communication Cycle, Macrotick and Microtick

To begin this new chapter about the aspects of synchronisation and Global Time during normal operation of communication, let us remind ourselves in a few words of the time hierarchy which is set up in FlexRay.

14.1 The FlexRay Time Hierarchy

Figure 14.1 is a reminder of the four principal levels of this: the communication cycle, the macroticks, the microticks (μTs) and the local clock.

14.1.1 Communication Cycle

Let it be said once and for all: in a FlexRay system, by definition the 'communication cycle' consists of a whole number of macroticks (MTs). The number of MTs per cycle is identical for all the nodes of the same group (cluster), and remains the same from one cycle to another. Also, at every instant, all the nodes must have the same cycle number and must manage it at the same time.

14.1.2 Macrotick

The purpose of 'macroticks' – MTs – is to set up a first relationship between the physical signal present on the network and the μTs.

The 'MT' is an interval of time concerning a particular set of participants of the network (called 'cluster-wide'). Essentially, it represents the smallest unit of Global Time (the finest time granularity) of the network.

The local duration of each of the MTs associated with a node consists of a whole number of μTs of this node, and as this chapter will show, at every instant, the value of the 'μTs per MT' ratio is calculated/adjusted/established locally using an algorithmic synchronisation procedure (see next page). Its form and structure are therefore not entirely

FlexRay and its Applications: Real Time Multiplexed Network, First Edition. Dominique Paret.
© 2012 John Wiley & Sons, Ltd. Published 2012 by John Wiley & Sons, Ltd.

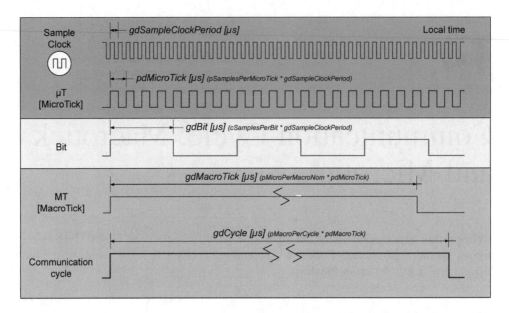

Figure 14.1

linked to a simple story of electronic mechanics as in the case of μTs, but are the result of a clever calculation.

Given that we indicated above that the values of μTs are specific to each node of the network, as a function of the frequency of its local oscillator and the values held in the configuration registers of its internal predividers, the values of the μT/MT ratio will also be different from node to node. Additionally, in the course of a communication cycle, the number of μTs per MT can be different from one MT to another, within the same node.

Although any one MT consists of a whole number of μTs, the mean duration of all the MTs of a whole communication cycle can be a non-integer value; that is, it may consist of a whole number of μTs plus a fraction of a μT. The purpose of these timing adjustments, provided by clever calculation of the value of the MTs – themselves directly linked to the μTs, which are tied to the frequency of the microcontrollers of the CPUs – is to provide time synchronisation between the signals present on the network and the microcontrollers.

14.2   Synchronisation in a Network of TDMA–FlexRay Type

Synchronisation of the nodes in a FlexRay network takes place twice, or at two levels, called 'macroticks' and 'microticks'.

14.2.1   Statement of the Problem and Requirements to be Satisfied

Figure 14.2 shows the problem to be solved very clearly. As you can see, several nodes are connected together somehow. How it is done is very variable, depending on what topologies are used.

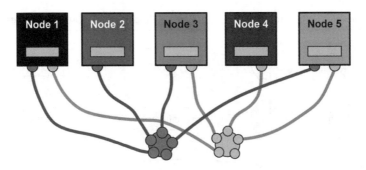

Figure 14.2

14.2.1.1   Statement of the Problem and its Constraints

Each node/participant of the network has:

- its own hardware, including a specific microcontroller dedicated to the task which is the responsibility of the node;
- its own local clock (with its own oscillator, driven by the appropriate quartz), which, for numerous reasons, has a different value from other nodes (for example linked to the requirements of the application for timing and (low) power consumption);
- after division of its local clock by the internal dividers of the microcontroller, its own local bit rate (expressed in bit/s) – which of course will be very close to what is wanted for the whole network, but very slightly different from that of the other nodes;
- thus its own local view of what should be, for it, the time duration of the communication cycle;
- and, at no extra cost, its own local instant (point) for starting the communication cycle.

There is no phase relationship between all the clocks of all the nodes. There are, therefore, phase offsets between all these rates. Normally, these phase offsets should be expressed in degrees of phase relative to an arbitrary starting origin of the cycle, but in our case, with a constant bit rate, they will be expressed as time (nanoseconds, microseconds and μTs).

Figure 14.3 illustrates the general, very normal problem of the relationships that can exist between the clocks.

Figure 14.4 presents, in a different general form, the concept of timing variability of local clocks of the CPUs in a FlexRay network.

In this figure, the abscissa represents the 'physical time' according to an external absolute time reference, and the ordinate shows how the time of each controller develops relative to the reference time. With the same units shown on the two axes, if all the clocks of the controllers on the network changed strictly at the same speed as that of the reference, all the straight lines representing the variations of the clocks of the different controllers should be straight lines at 45° – which is never the case, since some micro-controller clocks tend (. . . like all watches) to run fast or slow – and thus not to have the same 'rate'!

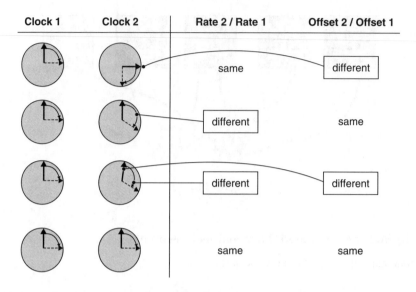

Clock 1	Clock 2	Rate 2 / Rate 1	Offset 2 / Offset 1
		same	different
		different	same
		different	different
		same	same

Figure 14.3

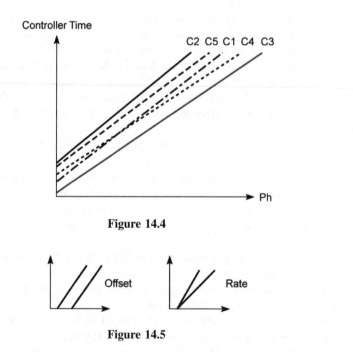

Figure 14.4

Figure 14.5

The two curves of Figure 14.5 show:

- on the one hand, that if their clocks change at the same rate (no tendency to run fast or slow), the indicated time value is not the same, and that therefore there are phase differences between them (offset);

- on the other hand, that with the same starting phase, one clock advances relative to the other, and that there is then a difference of speed (rate) between them.

14.2.2   Requirements to be Met

In parallel with the fundamental problem presented in the preceding paragraphs, let us now look quickly at the numerous other requirements to be met:

- it is necessary to be able to support great variability of the number of participants of the network (what are called, in pompous terms, the 'scalability' of the network and the topological 'variabilities')
  - on the one hand, on a single communication channel:
    * the presence of multiple nodes in a cluster,
    * the presence of multiple mutually independent clusters,
    * the fact that multiple clusters can be connected to each other,
    * and so on;
  - on the other hand, in a mode in which two communication channels function simultaneously:
    * cluster-to-cluster relationships, on channel A–B,
    * and so on;
- it is also necessary to have a very precise global clock (we want to, and we will, do everything so that the bit duration is equal, or as close as possible, to the desired ideal abstract value, in this case 100 ns);
- to satisfy reasonable industrial scenarios by achieving a maximum overall error (phase + bit rate) of 1 µs (that is, equivalent to 10 bits at a bit rate of 10 Mbit/s);
- to retain high-performance use of the bandwidth of the system;
- to arrange 'fault tolerance', so that up to two asymmetrical faults can be accepted;
- to have great intrinsic robustness, so as to have the ability to survive for several communication cycles without the aid of a synchronisation device;
- to tolerate the drifts of the quartz crystals in use being of the same type as those that are usually met in automotive industry applications;
- and so on.

The game is therefore to get everyone in step with each other, with the aid of a so-called synchronisation device and process, the purpose of which is to define or derive a common base time called 'Global Time' from the local clocks of each of the individual nodes on the same cluster,[1] the end purpose of which is to obtain (see Figure 14.6):

- a global clock for the cluster;
- a global/common startup time of the cycle for the cluster;
- a common cycle duration for the cluster.

---

[1] As a reminder, the generic term 'cluster' or 'group' means all those elements/nodes of the network which have more or less related functional properties and timing constraints to share. In general, a single network supports multiple clusters of nodes, for example on the FlexRay network which includes nodes A, B, C, D, ... S, cluster 1 can be formed by nodes A, C, F, G, P, cluster 2 by nodes B, E, M, N, and so on. So when cluster synchronisation is mentioned, it is group by group.

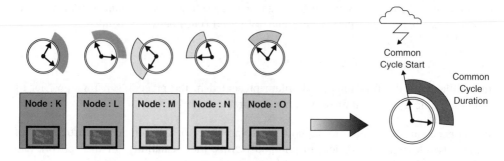

Figure 14.6

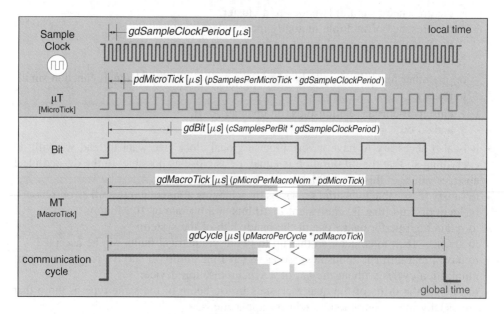

Figure 14.7

We will show below that, consequently, a local node knows the value of Global Time by referring to two parameters in particular: the cycle count and the MT count.

All of this is got into step in several stages, by juggling with the famous 'MTs' and 'μTs', and more specifically with the value of the ratio which links them, and which has already been discussed at great length!

Figure 14.7 shows the overall view of the principle of operation of FlexRay message synchronisation using MTs and μTs.

Now that the problem has been clearly stated . . .

14.3   Proposed Solution to the Problem

Because of the extent of the problems to be solved as described above, FlexRay's proposed solution for synchronisation between nodes is complex, and presenting it clearly for

teaching purposes is not easy. So to avoid difficulties of comprehension, we have chosen
to divide it into two parts, the first about the concept of 'synchronism', the second about
the field of 'isochronism', and at the end of the chapter, we will finish by establishing
the link between the two concepts.

> **NOTE**
>
> As a reminder, here are some definitions:
>
> - the English term 'time' corresponds to the Greek 'khronos';
> - isochronous, formed from the Greek prefix 'isos', equal, and khronos means 'of the same time', 'of
>   the same duration', 'of equal duration', as far as we are concerned, same period, same frequency, same
>   speed, same rate;
> - synchronous, from the Greek 'sunkhronos', the 'sun' part specifies that it 'is done at the same time',
>   'at the same instant', 'in phase', and for us that will be linked to the concept of offset.

14.3.1 Introduction: Forewarned is Almost Cured

To provide a possible slight timing adjustment of the duration of the communication cycle
in order to create a Global Time for the network, it was necessary to introduce, at the
end of the communication cycle, an interval of brief duration during which it is decided
to transmit nothing on the medium; this is the network idle time (NIT). Not too long, not
too short, but just right to be able to shorten or lengthen it as required ... and to do lots
of other things.

14.3.2 Description of the Chosen Method of Ensuring Time Synchronisation of the Nodes of the Network

So that the chosen solution performs well, the corrections of speed and phase must be
carried out using the same methods on all the participants of the network. To do this,
let us begin by stating the chosen method and the operating principle which are used to
create a common Global Time for all the participants of the same cluster.

The chosen method takes place in several successive stages, the most important of
which are as follows:

- determining the nodes which participate in a synchronisation sequence;
- divergence measuring stage;
- stage of calculations and determining the corrective values;
- and finally, applying the corrective values to the participants of the system.

Let us now look more carefully at each of these stages one by one.

14.3.2.1 Determining the Participants in a Synchronisation Sequence

Firstly, one defines/chooses the nodes (a cluster under consideration) which one wishes
to see participating in a synchronisation sequence of their respective clocks. The very
impersonal 'one' in the previous sentence can hide numerous possibilities. It may mean
the system architect, it may be the task that the node carries out and the particular

Sync Frame Scheduling								
Node	Static segment							
	Slot d	Slot h	Slot m	Slot o	Slot r	Slot u	Slot y	Slot z
K	Syncframe							
L								Syncframe
M			Syncframe					
N						Syncframe		
O		Syncframe						
P				Syncframe				
Q					Syncframe			
R							Syncframe	

Figure 14.8

conditions; in short, very numerous reasons push a node into wanting to participate in a synchronisation sequence.

14.3.2.2 Indicating the Determination of a Node to Participate in the Sequence

To signal that it wishes to participate in this sequence, during a communication cycle, during and at the start of the slot of the static segment which is assigned to it, the transmitting node under consideration, via the fourth bit of the header of the communication frame of the slot under consideration, indicates to all the other participants of the cluster the fact that the content of the frame which it is about to transmit is a so-called synchronisation frame, which acts to participate in the synchronisation of the network. Consequently, the other participants wishing to participate in this work phase will be obliged to send synchronisation frames in the slots which are specific to them in the static segment.

To sum up this part of the chapter, Figure 14.8 shows the example of a communication cycle during which certain nodes on the network have wished to participate in the synchronisation phase by sending a synchronisation frame during the slots assigned to them.

COMMENT

Some specific comments concerning the form of sequences and synchronisation frames:

- The maximum number of nodes authorised to participate in the synchronisation sequence is fixed at 15.
- Even if a single node is authorised to occupy multiple slots in the same communication cycle, this node can send only:
  - one synchronisation frame per slot;
  - one synchronisation frame per cycle.
- With the exception of the active node in the course of transmission, all the nodes of the network measure and calculate the divergences of offset ('phase') and bit rate relative to all the other nodes which are present during this communication cycle.
- In the case of FlexRay systems operating on two communication channels, normally the synchronisation frames are sent simultaneously on the two channels.

14.3.2.3   Measurement Methods and Measurements

The basis of the measurement method explained below is the fact that, at the time of the above sequence, seen from its own window, each of the nodes on the network, for the duration of the static segment, observes and measures all the actions of the other nodes which are also present on the network and have declared their wish to participate in the synchronisation sequence.

Let us now look more specifically at how all this minor espionage works.

All the local observations (at each of the nodes which are present at the time of the synchronisation sequence) can be carried out only by using their local weapons, which consist exclusively of the knowledge of the values of their own local time references; that is, their own local clocks. Each of them therefore knows precisely its characteristics (in terms of related frequency and phase values) and also the values of the parameters coming directly from them. Each of them is therefore capable of quantifying, in values of its local μTs, all the timing divergences that it is likely to observe – apart from the precision or uncertainty of their respective timing measurement granularities, which are strictly due to the presence of time durations of finite values of each of their respective μTs.

Also, during the static segment, while these observations are being made, the attitude of each of the nodes in the observation state is summarised in two things:

- hoping that, in relation to its local clock, something forecast occurs at the hoped-for instant (this is why using the static segment was chosen, since in it access to the network is deterministic!);
- noting in a scratchpad, using and by reference to its own clock (the only one available to it), the instant when the hoped-for event actually occurs.

Next, to place itself in relation to its fellows, all each node has to do is to draw its own conclusions locally between the hoped-for moments when something should have happened and those when something actually happened!

It is now necessary to reflect, to determine what one has to hope for on the one hand, and observe on the other hand, in order to achieve 'synchronisation' and 'iso-synchronisation' between nodes.

14.3.2.4   The Iso-Synchronisation Aspect; that is, about the Simultaneous Start of All the Nodes, so the OFFSET Concept

Finally, when all the participants of the network are at the same speed, the important point will be to make all the frame transmissions of each node start at the right instant. On principle, by measuring the actual instant at which they start relative to its own hoped-for point, the local node can draw conclusions directly about its own relative offset.

14.3.2.5   Measuring the Time Offset

A new communication cycle starts on the medium:

- In the light of its own local clock, so of its own local view of Global Time, when it has reached the predefined values (action point, and so on), a node A which needs to communicate, without further ado, starts transmission of its message (frame) in the static slot reserved for its use.

- Additionally, another node Y on the network waits to receive – at a hoped-for instant linked to the local value of its own view of Global Time – the transmission of the message (frame) from A. Obviously, it's either early or late, otherwise it wouldn't be interesting! Node Y, using its own clock, notes the instant at which the frame arrives, and measures (by counting the number of its μTs), at Y, the time difference (in relative values, + and –) between 'the actual instant of the arrival of the start of the message from transmitter A' and 'the hoped-for instant of the arrival at Y of the message from A' (this value being calculated, for example, from the identifier of the message, or by using predefined messages such as 'sync frames'). The controller of node Y can measure, evaluate and quantify (with the finest granularity of local measurement of the node, which is the time resolution of the μT) the 'Global Time distance' which separates it from the controller of node A (see Figure 14.9).

IMPORTANT COMMENT

Since each of the nodes of the network (B, C, D, and so on) is different from its fellows, and node Y, as its only reference, has nothing but its own clock, it repeats this procedure in each of the other static slots taking part in the synchronisation sequence. During these static slots, all the other controllers communicate on the network during the communication cycles (see Figure 14.10). Node Y encodes all the obtained results (the offsets and different phase divergences which exist between itself and each of the various participants on the network and in the static segment) as signed algebraic numbers (positive and negative) of μTs (its own).

Figure 14.10 shows an example of the result of these measurements, with four nodes participating in the synchronisation sequence.

The phase divergences shown in this figure give the overall image of the network at a given instant. Of course, each node on the network has its own view of this overall representation, for example in the case above:

- for controller **C1**, at **C1** C2 C3 C4 respectively, we have:
  0, −2, −6, −9 µT;
- for controller **C3**, at C1 C2 **C3** C4 respectively, we have:
  6, +4, **0**, −3 µT;
- for controller **Cn**, at C1 C2 C3 C4 respectively, we have:
  …… …… µT

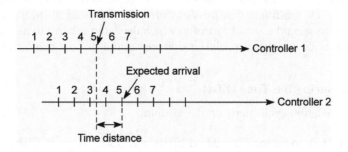

Figure 14.9

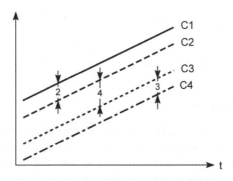

Figure 14.10

Node X		
Slot	Measured Values (µ Tick s)	
	Offset	
d	4	
h	−3	
m	7	
o	6	
r	−1	
u	1	
y	0	
z	−1	

Measurement Values are relative to the local timing of node

Figure 14.11

For better understanding, let us go back to our example (Figure 14.8). Figure 14.11 shows all the measurements of offset divergences.

14.3.2.6   Measured Values

Additionally, because of the repetitiveness of FlexRay communication cycles, these measurements can be carried out at the start of all the frames included in the same slot in the course of the succession of different cycles.

14.3.2.7   For Synchronisation, so Everything that Concerns Comparisons of Speeds and the Rate Concept

For a 'local' node, what matters is to measure the divergence between its own bit rate, for example the duration which it would itself assign to its communication cycle via its own local clock following configuration of its own registers, and the bit rate of one or more nodes on the network. Here, too, to be able to correct itself later, the principle which is adopted consists of the node observing the respective bit rates of the other nodes and comparing its own bit rate with the others.

14.3.2.8 Measuring Duration

To carry out this new stage concerning measurement of the duration or periodicity of repetition of the same frame identifier from one cycle to another, the adopted principle of measurement is very simple. Because the cycles and their structures are repetitive, by using its own clock and therefore its own local μTs, a particular node of the network measures the time lapse (duration, length of time) which exists between two moments which represent in time the repetition which exists between two events with the same meaning, for example the duration between the starts of the same static frame from one cycle to another (for example, the start of the frame of slot K of one cycle and the start of the frame of the same slot K of the following cycle). In other words, to carry out the procedure for speed correction, it is always necessary to consider what happens on a pair of communication cycles, taking as the reference (start) of this pair the even-numbered cycle, and not on a single cycle as was possible for measuring the offset.

Figure 14.12 illustrates this measurement method.

At this stage, the local node, which knows exactly the number of μTs which it has itself assigned to the duration of its cycle, and having now measured, according to its own μTs,

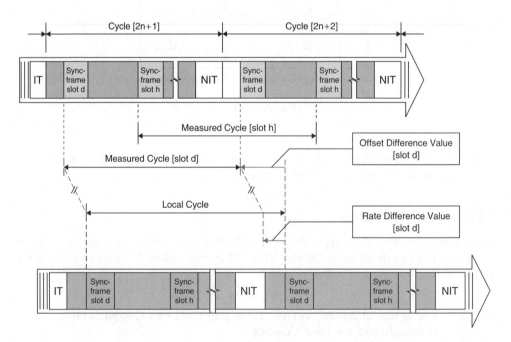

Figure 14.12

Node X		
Slot	Measured Values (μ Tick s)	
		Rate
d		−9
h		1
m		−8
o		2
r		12
u		6
y		−9
z		−2

Measurement Values are relative to the local timing of node

Figure 14.13

the value of the duration of a cycle of another node, is entirely capable of deducing from it its divergence of speed relative to the transmitting node under consideration, simply by calculating the signed difference (+ or −) between these two values.

Figure 14.13 shows an example of a table of divergences, calculated following these series of measurements, which are carried out during the static segment and synchronisation phase on the other nodes on the network, and seen from a specific node.

14.3.2.9 Measured Values

The two types of measurement described in the preceding paragraphs are therefore carried out throughout the duration of the static segment, where the static slots are.

In fact, these two types of measurement – duration (rate) and phase (offset) – are carried out in a single pass, and the housekeeping is done afterwards, as we will now explain.

14.3.3 'All in One' Measurement

Each node of the cluster then measures, using its own clock (because it's the only one it has), the duration in local μTs of the communication cycles of each of the nodes which it receives, over a pair of cycles, by measuring the time difference which exists when instances of the start of the same synchronisation message sent in two successive frames are received (for example synchronisation frame D of slot Y of the first cycle to the same synchronisation frame D of slot Y of the second cycle during the static segment). This measurement enables the controller at node K, transmitting in slot X, using the timing of its own local clock, to estimate the duration of the cycle of the controller of node Y (according to its own time base X, its μT counter, MT counter, slot counter and cycle counter).

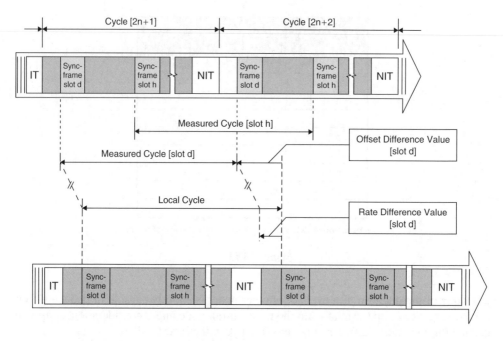

Figure 14.14

Figure 14.14 illustrates this phase of operation, with:

- at the top of the figure, the cycles circulating on the network, and the frames transmitted in the static slots by all the nodes, based on the 'cluster time' in the course of use;
- at the bottom of the figure, the local view of the same communication cycle circulating on the network when it arrives at the terminals of one of the specific nodes of the same network.

Several conclusions can be drawn by the local receiving node:

- the local receiving node, having initiated its measurement (using its local clock), when it receives a start of a particular frame belonging to a precise slot, is capable of calculating for itself (seen from its window) at what moment 'normally' the start of the repetition of this same frame of the same slot should appear in front of it during the next communication cycle;
- because of the differences of bit rate or phase between the slot of the transmitting node and itself, this second start of frame of the second cycle does not appear at the hoped-for instant, and the local receiving node is then capable of evaluating (using its local clock and its μT counter) the difference, that is the positive or negative local delta μT, between the moment it hoped for and the sombre reality of what happens at its terminals.

This measurement enables the receiving node, using its local clock, to calculate two quantities:

- On the one hand, seen from its receiving window, the elapsed time for which it has waited for the received frame during a specific slot, and thus to conclude locally whether its own local cycle time was too long or too short, so that it can then shorten or lengthen it, adding to or subtracting from the number of local μTs forming the MT. By doing that, the receiving node adjusts its bit rates to the transmitting node. In Figure 14.14, the 'rate difference value' illustrates this.
- On the other hand, the arrival time of the frame, which is offset in time (propagation time of the medium, presence of active stars, and so on) and by the local clock of the node under consideration.

Figure 14.14 emphasises (counted in μTs at the local level of the receiving node):

- the differential values of time offsets at the start of cycles;
- the differential values of bit rates – cycle duration/frequency.

The same thing happens for each of the controllers on the network X, to know what happens for the controllers of the slots Z, and so on throughout the length of the static slots of the dynamic segment.

Summarising, for the controller of node X it is sufficient to measure all the differences relative to the other cycle lengths Y, Z, and so on.

> **COMMENT**
>
> At each of the controllers of each of the nodes arranged on the network, the smallest unit – the resolution, the granularity – of measurement of time differences equals the duration of their own local μT.

14.3.4   Calculating the Corrective Values of Offset and Rate

Now that each of the controllers on the FlexRay network has (with the aid of measurements which are carried out and collected within a scratchpad memory organised in the form of two tables, respectively Offset and Rate) all the information about time differences, measured divergences and values (in local μTs), we can now describe the mechanism for calculating the values (whole numbers of local μTs) which are used to correct the offsets (phase) and bit rates.

For this purpose, each node which has participated in the synchronisation sequence will:

- first, execute a 'thresholding' procedure on the values of the divergences (using a specific algorithm called 'fault-tolerant midpoint (FTM)', described below);
- then, calculate and deduce what are the corrective values of offset and cycle duration which the node must apply to itself relative to the value of the most reasonable/suitable Global Time for all the nodes of the network which it has just estimated.

14.3.4.1   Principle of the 'Fault-Tolerant Midpoint (FTM)' Thresholding Algorithm

To calculate the values to be used to correct the bit rate and phase (offset) of the parameters of the communication cycle, FlexRay lays down the use of a particular algorithm called

FTM Algorithm	
Number of Sync Nodes	Parameter k
1 - 2	0
3 - 7	1
> 7	2

Figure 14.15

'fault-tolerant midpoint'. The architecture of this algorithm is not new, and actually dates from the first conventional techniques of 'averaging'. The purpose and effect of using it is to minimise, on the one hand, the hardware designs of the components and, on the other hand, most of the effects due to certain erratic faults which can occur, but unfortunately it still preserves some systematic second order errors (run time jitter, granularity limit and non-linearity of oscillators).

The functional principle of the FTM algorithm is as follows. Once the measurements have been carried out and the two summary tables of divergences have been constructed, each controller on board each node:

- arranges in descending algebraic sequence, each on their side, the (signed) values of the divergences measured in local μTs corresponding respectively to the 'offset/phase' and 'bit rate' headings;
- then removes the most extreme values from these measured divergences (which may be due to measurement errors, for example). Before going further, it should be noted that the FTM algorithm implies taking account of the number of nodes which are actually present during the synchronisation sequence (with a maximum of 15 participants during the procedure). In fact, the number of extreme values which are removed depends on the number of nodes which are present on the network during the synchronisation phase of the latter. The reason is simple: the more nodes there are, the more chances there are of extreme values far from the mean value. The table of Figure 14.15 indicates in the form of the 'k' parameter the number of extreme values that must be removed, as a function of the number of nodes in the network cluster under consideration;
- now the 'k' greatest and smallest measured values are removed;
- next a specific 'algebraic mean' is calculated as the signed sum of the extreme terms of the remaining table (that is, the sum of only the greatest AND the smallest of the remaining values), and this is then divided by 2;
- if the result of this division is not an integer, the obtained value is rounded to the nearest low value, 'r';
- the thus-obtained result gives the corrective value which the local node will have the aim of applying.

To illustrate the whole of this phase of the procedure, let us take an example.

14.3.4.2   Example of Calculation of the Corrective Values of Offset and Bit Rate

Figure 14.16 summarises and generalises the examples of values of divergences/ differences of 'offset' and 'rate' presented above, of course expressed in local μTs, for example at node 'X' as a function of the different frames received in the static slots where the synchronisation frames associated with the different controllers are transmitted.

As indicated above, using the FTM algorithm, the procedure is identical on the two columns of Figure 14.16, on the one hand for the 'rate', on the other hand for the

Node X		
Slot	Measured Values (μ Tick s)	
	Offset	Rate
d	4	−9
h	−3	1
m	7	−8
o	6	2
r	−1	12
u	1	6
y	0	−9
z	−1	−2

Measurement Values are relative to the local timing of node

Figure 14.16

FTM Algorithm for Rate Correction of node X				
Rate	Sorted	Selected	Sum	Midpoint
−9	12			
1	6			
−8	2	2		
2	1	1		
12	−2	−2	→ −6 / 2	→ −3
6	−8	−8		
−9	−9			
−2	−9			

Figure 14.17

FTM Algorithm for Offset Correction of node X				
Offset	Sorted	Selected	Sum	Midpoint
4	7			
−3	6			
7	4	4		
6	1	1		
−1	0	0	→ 3 : 2	→ 1
1	−1	−1		
0	−1			
−1	−3			

Figure 14.18

'offset'. As Figures 14.17 and 14.18 show respectively for the rate and for the offset, the values are:

- reordered/sorted in descending algebraic order;
- thresholded (in the example, apart from node 'X', eight other nodes are present during the synchronisation phase, so at least two values must be removed at the extremities);
- the two remaining extreme values of the table are then summed algebraically and averaged, to obtain the corrective value.

14.4   Application and Implementation of Corrective Values

Now that it knows the corrective values to be applied to the 'offset' and 'rate', the node under consideration comes to the final stage of the synchronisation sequence, which consists of applying the correctives values above.

Let us now look at how and where these corrective values will be applied.

The node under consideration (like all its other fellows) has two strong constraints:

- on the one hand, at a given instant, its own local clock[2] is what it is, in initial value and drift;
- on the other hand, the network requires that the communication cycle is composed of k MTs – which is 'the' constant of the network.

14.4.1   Offset and Offset Correction

14.4.1.1   Stating the Problem

First, let us assume that all the nodes function strictly at the same bit rate. At this point in the chapter we do not yet know how we have succeeded in bringing all the nodes to the same bit rate, but for the moment let us make this assumption.

---

[2] To its own 'clock', according to the very strict, well-defined rules according to whether correction of offset (phase) or correction of the value of bit rate (frequency) is involved. We will examine these in a few paragraphs.

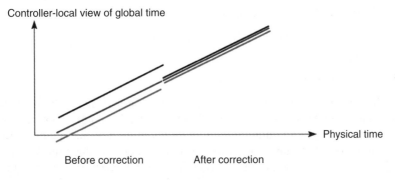

Figure 14.19

14.4.1.2   Principle of Correction

First we will look at offset correction, the purpose of which is to reduce the phase errors which can occur between oscillators with identical frequency values.

14.4.1.3   Application of Offset Correction

Finally, at the start of each second cycle (odd), each node adjusts (upward or downward) its own view of Global Time, using the calculated corrective offset term (in μTs). Figure 14.19 shows the states before and after offset correction.

After this algorithmic processing of offset corrections, some systematic phase errors remain, due to:

- run time;
- the resolution/granularity of the measurement quantified in μTs;
- divergences of bit rates between controllers (see following paragraphs);
- higher order phenomena such as effects of non-linearities of the oscillators.

To function correctly, offset correction requires that the bit rates are quite close. Unfortunately, in an industrial context such as that of the car, over a period of 10 years drifts/ageing of quartz oscillator frequencies by ±250 ppm can be observed. When we add that it is necessary to take account of a safety factor of a few thousands of ppm, it turns out that we have to allow for a value of the order of 2000 ppm. Apart from the fact that offset correction cannot function correctly, for a system operating at 10 Mbit/s that can result in a variation of 40 µs (a slippage of 400 bits) on a communication cycle of 20 ms – which in terms of number of bits is enormous! Given that the aim for a distributed system of TDMA type is that its synchronisation device should be effective, in the case of FlexRay it is necessary to succeed in holding the microsecond (that is, 10 bits of 100 ns over 20 ms), and what we have just explained shows that offset correction is not enough, and that as well as implementing it, we must also consider implementing bit rate correction.

14.4.2   Rate and Rate Correction

Let us now go on to describe the rate correction device.

14.4.2.1   Principle of Rate Correction

To make the bit rate values identical for each of the participants of the network – what everyone pompously calls 'synchronisation', and strictly speaking is not quite that – we will act so as to bring all the time lengths of cycles of each to an equal quantity. To do that, using the quantified rate error correction information, the node under consideration modifies/corrects the time length of its cycle, in the same spirit as what was done in the sections about phase correction – but of course slightly differently, otherwise it would be too simple.

By application of the principle stated above, the variations of run time will be annihilated. In contrast, as we have shown up to now, offset correction influences measurement directly and systematically.

14.4.2.2   Conclusion

In principle, a rate corrective value can be calculated, determined and indicated only every two cycles, and it is therefore also necessary to correct the offset only every two cycles, and for that it is necessary to distinguish even and odd communication cycles. As well as providing a continuity index function, this is one of the reasons that communication cycles are numbered, and it is then easy to find the last digit of the communication cycle counter, to know immediately what is the parity of the communication cycle in the course of execution.

Let us go back to the chronology of all these operations in detail.

14.4.3   Where, When, How to Apply the Corrections?

As we have indicated several times, the FlexRay communication cycle includes four segments, of which two are obligatory, the static and the NIT, and two are optional, the dynamic and the symbol window (SW).

14.4.3.1   Where

In principle, all the measurements of offsets and rates can be carried out only during the deterministic part of the communication, so only during and throughout the duration of the static segment – which fortunately is obligatory! All the readjustments of the parameters that are associated with the phases and bit rates, and are in the registers of the microcontrollers of the nodes of the network, can be carried out – as quickly as possible – only during the moment of calm which precedes the new tempest of the following cycle – that is, during the NIT – which is why it is there! It is therefore during the NIT – having, just in case, the last little value obtained on the last static slot – that it is possible to calculate, do the thresholding, and so on, to be able, during the NIT, to apply corrections to the parameters of the microcontroller.

We have just thoroughly wrung the neck of Where, but not When or How!

14.4.3.2   When

We have seen that the offset measurements can be and are easily carried out cycle by cycle – so every cycle – but that those associated with measurements of the duration of cycles can, in principle, only be carried out on a pair of cycles. That leads to the conclusion that to have good consistency of bit rate and phase, the modified values of the microcontroller parameters will be inserted only every two cycles, when the corrective values of rate and offset are available simultaneously, and that by doing this, in principle these corrections will act on the following pair of cycles and not from cycle to cycle, since, on the one hand, the measurements are carried out on a systematic pair of cycles, 'even/odd', 'even/odd', for example (2/3, 4/5, and so on), and not on sliding pairs, 'even/odd', 'odd/even', 'even/odd', for example (2/3, 3/4, 4/5, and so on), and on the other hand, modifying rate without offset and vice versa is quite ridiculous.

So that's When sorted out! Now let's go on to How!

14.4.3.3   How

Correcting Bit Rate and Offset

Generally, the signed corrective values above are added to or subtracted from the value 'd'[3] of the representation of MTs, to the microcontroller's own cycle length, after it has done the measurements. This value is used to adjust (increase or decrease) the length of its communication cycle sufficiently.

Correcting Offset

- Having decided to work with 'even/odd' cycle pairs, the offset correction is initiated, carried out and takes effect only during the NIT of odd cycles.
- By a principle of the FlexRay protocol, since the cycle comprises a constant number of MTs, the position of the start point of the new communication cycle can then be synchronised/adjusted (forward or backward) to the Global Time of the network by lengthening or shortening the duration of the MTs contained in the NIT.

Figure 14.20 illustrates this.

---

[3] We refer you to Part B concerning the parameter 'd', when we explained the relationships which existed between μTs and MTs – all the great mysteries will be explained like that one day!

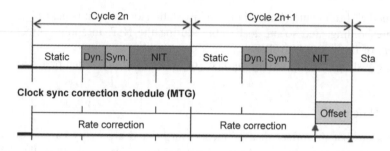

Figure 14.20

Correcting Frequency/Bit Rate

- The duration of the communication cycle is adjusted at the local level of the node under consideration (lengthened or shortened).
- By a principle of the FlexRay protocol, since the cycle must always comprise a constant number of MTs, a local node can modify the value which it gives to the duration of its cycle only by modifying the duration of the MTs that form it – and therefore the number of μTs that it assigns to an MT – and therefore act on the value of the ratio (μT/MT). All that seems very simple, but to homogenise and smooth the durations of the MTs over the duration of the cycle (in fact over two consecutive cycles, 'even/odd', since corrections can only be done on two cycles), the number of μTs corresponding to the corrective value must be such that the distribution of the number of μTs per MT is adjusted and distributed over all the even and odd communication cycles.

Figure 14.21 shows well the instant at which the rate correction is carried out – that is at the end of the even/odd cycle pair – and also shows that the corrective action is carried out during the following even/odd cycle pair, and so on.

14.5 Summary

After all the actions of these two corrections, all the bit rates and phases of all the participants of a cluster are 'synchronised', and consequently all the nodes, each at their own

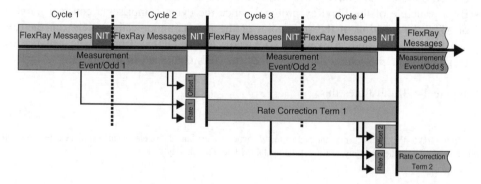

Figure 14.21

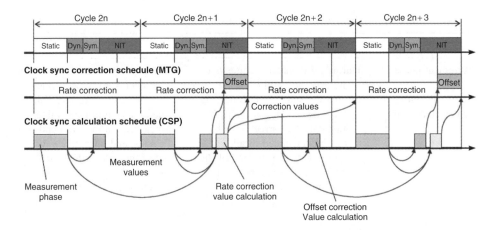

Figure 14.22

front doors, have invented 'one unified, uniform Global Time' for the whole network – for the next two communication cycles! Obviously, these operations are repeated again and again, and the network is constantly being synchronised.

Summarising, if you have followed the course of this chapter well, you will have understood that the NIT was put there specifically so that it can provide a time slot, the shortest possible, so that play can be taken up, and that although nothing happens visually (according to the oscilloscope) on the network, all the processors of the nodes work intensely during the NIT to construct this totally abstract Global Time, of which no-one is able to say the true value at any instant, except that it is made so that the bit rate of the network is very, very, very close to 10 Mbit/s, and therefore that the bit duration is 100 ns!

Figure 14.22 illustrates and summarises the whole of the stage of adjusting the values of phase and bit rate. On the one hand, it indicates the instants during which the divergences of duration (rate) and phase (offset) between signals on the network and in the nodes are measured and calculated, and also the moments at which the time values of the time slots are corrected, so that the timing of the whole network is consistent.

In this figure, it is important to note:

- that the corrective values of phase are calculated every cycle;
- that the corrective values of bit rate are calculated every two cycles;
- that the corrective values of rate and offset are applied only every two cycles;
- that the offset correction benefits from the duration of the NIT to offset, to a greater or lesser extent, the start of the following cycle, and its action is maintained on the following cycle;
- that the rate correction is applied identically to the two following cycles.

14.5.1 Supplementary Note: Example of Time Hierarchy

To be more specific, this brief technical appendix gives an example, of which Figures 14.23 and 14.24 show the particular chosen time hierarchy.

Table 14.1   Sum up of offset and rate measurements

	Offset	Rate
Measurement phase	During all the static segments of the even and odd cycles	During all the static segments of the cycles, over two cycles
Phase of calculating the corrective value	So that the start of the following cycle takes place at the right moment, the corrective values will be calculated in all cycles, between the end of the static segment and before the start of the NIT	Calculation of corrective values takes account of the values measured in the even and odd cycles, and takes place just after the end of the static segment of the odd cycles (so one cycle in two)
Phase of applying the corrections		
When	Only one cycle in two, on the odd cycle	Every two cycles, but just before an even cycle starts
Where (place)	Only during the 'offset correction segment' of the NITs (between cycles), and finishes before the start of the following cycle	Over the full extent of the communication cycle
How	Phase correction (offsetting the start of the new cycle) is carried out by adding or subtracting μTs	Corrections of the value of the bit rate are carried out/distributed using a whole number of μTs distributed in the MTs

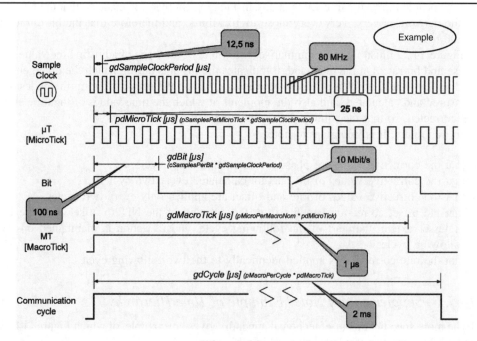

Figure 14.23

gdSampleClockPeriod	12,5	ns
pSamplesPerMicroTick	2	1/µT
pdMicroTick	25	ns
Bus Speed	10	Mbit/s
gdBit	100	ns
MicroPerBit	4	µT/Bit
cSamplesPerBit	8	-
gdMacroTick	1	µs
pMicroPerMacroNom	40	µT/MT
gMacroPerCycle	5 000	MT
pMicroPerCycle	200 000	µT
gdCycle	5 000	µs

Figure 14.24

In the case of readjustment of the Global Time of the network, always based on 5000 MTs per cycle, it is the μT/MT ratio that will change. Let's go into detail about that.

Let us assume that at the end of five years, the drift/ageing of the frequency of the quartz oscillator of this node, which initially worked at 10 MHz, is +200 ppm (or 2 kHz). Such a situation, with all the values of the dividers precisely indicated in the table of Figure 14.24, causes a new communication cycle time of 5 ms = 5 000 000 + 1000 ns = 5.001 ms to be constructed at this specific node using the new values of μTs (25 ns + 5 ps), instead of the initially intended 5 000 000 ns exactly.

We can then have two different views according to the nodes on the network:

- for all the fellows of this node on the network, if they are still set at 10 MHz exactly, its cycle lasts 1000 ns longer, so for them the equivalent of exactly 10 bits more, or an equivalent divergence strictly equal to 40 µTs;
- for this specific node, the μTs of which are 25.005 ns, the divergence which it measures is 1000/25.005 = 39.99 of its own μTs – which for it is also 40 µTs, because of the granularity of the measurement – but the difference has been noted!

It is thus these 40 µTs which must be distributed over all the 5000 MTs.

RELAXATION – SMALL MUSICAL ANALOGY

After all these very serious pages, as entertainment, we offer you a 'musical' analogy to read.

To illustrate this concept of Global Time and the synchronisation which it underlies, let us take the example of a jazz quartet going onto a concert stage to interpret a musical piece.

Let us begin by setting the scene:

- the musical work which is presented is actually a suite of solo instrumental bars, and there is a passage in which all the instruments play together;
- all the musicians, using their own metronomes, have practised their own sections, on which is written the tempo at which the work should be interpreted, for example a rhythm of 60 crotchets per minute, for hours at home;
- a jazz quartet usually has no conductor, and a metronome is not usually available at the centre of the stage to provide an absolute time reference for everyone!

Although they are very used to playing together, all the musicians have their own ideas about what 60 crotchets per minute represent. So the piece begins:

- the first musician starts to play solo, and plays several bars, believing that he is playing at 60 crotchets per minute – but in fact, measured using an absolute time reference, he is playing at 59;
- the second, being excited when she starts, continues with her solo, also playing for several bars – at 62;
- the third continues at 60;
- the fourth at 61.

Finally the chorus, the common part. What happens? All of them have felt, for one or two bars, the small time difference which they have locally with each of their partners. Unconsciously, all of them, in their heads, while continuing to play, listen to the others, measure locally the divergence seen from their own windows (ear + brain), and correct themselves, to arrive at an appropriate resetting for the whole group. The first musician feels that he is too slow, the second and fourth musicians feel that they are much too fast, and so on. They all correct themselves dynamically in the space of one or two bars – so that the whole group now plays together at, for example, 60.3 crotchets per minute.

This value of 60.3 crotchets per minute represents the Global Time of this cluster (group), which is actually nothing but a joyous, absolutely unreal fantasy, since the hoped-for aim was exactly 60, but was completely unachievable because the four musicians were autonomous at the start of the interpretation (each with their local clock), and worked on the fly to synchronise themselves.

AUTHOR'S NOTE

Having also practised a lot of music in the course of my life, for information, you should know that the famous '... three ... four' at the start, as an introduction, can help but does not necessarily sort things out ☺!

15

Network Wakeup, Network Startup and Error Management

This chapter mentions numerous points which are often passed over in silence in general presentations of the FlexRay protocol. In fact, before and after everything works well, as indicated in the previous chapters, the network has to start some day or other, or better, each time one switches on one's vehicle, and afterwards it has to work correctly even if there are small problems! In short, here come wakeup, startup, error management and the rest.

15.1   Network Wakeup Phase

The paragraphs which follow will describe briefly the wakeup process of the network and the principal state diagrams related to it. But to begin with, we shall first consider that all the ECUs are placed in sleep mode to economise on energy, and that if an external event occurs, it will be capable of waking up the cluster.

For each of the nodes of a cluster, the wakeup phase refers simultaneously on the one hand to going from the 'power off' state to the 'power on' state, and on the other hand to going into the 'ready' state.

15.1.1   Node Wakeup Procedure

The wakeup procedure progresses according to the following scheme, which includes the succession of two different types of wakeup:

- **Local wakeup** – A wakeup is called a 'local wakeup' when it is due to the fact that a signal applied to a node via a separate wakeup input wakes up only this node. Now that this node has been woken up, it can become capable (if this is part of its task) of waking up the rest of the nodes of the cluster.
- **Global wakeup** – The node which is responsible for waking up the cluster then sends a particular signal called 'wakeup pattern (WUP)' (see description below) on the lines, to wake up the rest of the nodes of the cluster via the bus. This is what is called a 'global wakeup'.

FlexRay and its Applications: Real Time Multiplexed Network, First Edition. Dominique Paret.
© 2012 John Wiley & Sons, Ltd. Published 2012 by John Wiley & Sons, Ltd.

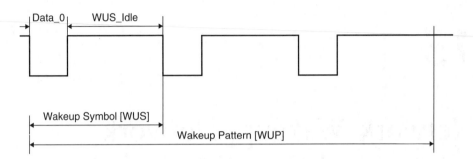

Figure 15.1

One obvious fact among others: For the global wakeup to take place, it is obviously necessary that before the WUP is sent, on the one hand all the line drivers of the other nodes of the cluster are powered, and on the other hand they are capable of waking up the rest of the components of their own nodes.

It should be noted that there is also a procedure called 'safety-related wakeup', during which some additional precautions are added to those described above (listening times, timeouts, and so on), to make the network wakeup phase secure.

15.1.2   Wakeup Frame – Wakeup Pattern – WUP

The frame called 'wakeup pattern', which is used to inform the network of a request for global wakeup, consists of a repetition of a symbol called 'wakeup symbol (WUS)', with 2–63 WUSs per WUP.

The symbol 'WUS' itself consists, for 4–6 µs, of a configurable number of Data_0 bits, and for 4–18 µs of a number of Idle bits.

Figure 15.1 shows this frame.

It is obligatory that all the FlexRay nodes support and can recognise the signal symbol called 'wakeup symbol', so that they can be woken up.

An example of the wakeup phase is given in Figure 15.2.

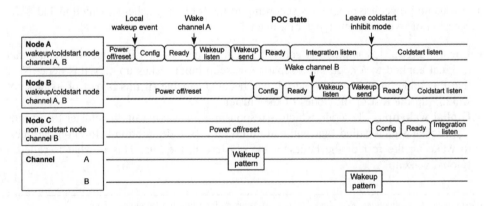

Figure 15.2

To conclude:

- The two successive phases we have described in the preceding paragraphs show that a single host node can be capable of managing the whole wakeup phase of a network (or a cluster), and that all the nodes are then powered (power on), woken up (wakeup) and ready to function.
- It should be noted at this stage that no data communication has taken place on the network (apart from the wakeup frame), and the nodes are not yet synchronised with each other.

Let us now go on to the next phase, that of starting up the network.

15.2   Network Startup Phase

Although all the nodes have slightly different local clocks, it is well understood implicitly that all are capable of producing naturally bit rates very close to 10 Mbit/s, but – as we have just said – after the wakeup phase, the nodes are still not synchronised with each other, and the sequencing tables of the communication cycles are not yet in place. Now, as you already know, the particular mode of access to the network of TDMA type used for FlexRay necessitates that all the nodes are strictly synchronised in the cluster.

The aim of the startup procedure is therefore to initialise, to implement the synchronism and to establish a common global clock, in order to produce common sequencing for all the nodes.

Again, obviously all the nodes must be powered and woken up, and when the system was designed, so offline, the network architect decided that some of them will be in charge and participate in this phase of starting up the cluster. They are called 'coldstart' nodes. For the moment, they wait, and subsequently they alone will be authorised to transmit synchronisation frames to try to initialise a startup sequence. Also, one of these coldstart nodes must be chosen/nominated to dominate (at least at the start) the cluster with its unsynchronised local clock. This is the leading coldstart node.

Now that the actors have been introduced, the startup sequence can begin.

- The leading coldstart node first begins by listening to/observing (coldstart listen) the network, to verify that there is no activity on it. Then, it first sends a collision avoidance symbol (CAS) of 30 Data_0 bits, the structure of which is equivalent to the MTS which was described with the NIT symbol. Its purpose is to inform the other participants of the network that there is a leading coldstart.
- The leading coldstart node then sends, on the two FlexRay channels A and B, for four consecutive cycles, the first of the startup frames (these are merely normal data frames dedicated to synchronisation with the 'sync frame indicator' and 'startup frame indicator' bits correctly set in the header of the frame, and therefore containing the definition of slot timings, and so on).
- Because, on the one hand, communication is now in place, and on the other hand, the first synchronisation frames from the leading coldstart node are sent and detected, the other coldstart nodes, after listening sensibly without noise for a minimum of four

cycles (and during this time having slyly begun adjusting the local clocks (rates)), try
to complete the startup phase of the network, and can themselves begin to send their
sync frames, to initialise and begin the full synchronisation of the cluster.

• This is followed by the integration of all the other nodes which are not coldstart nodes,
and which, apart from a few small details, follow the classic synchronisation rules
as stated in the previous chapter. This requires at least two startup frames from two
different nodes of the cluster. Once it is synchronised, a non-coldstart node can transmit
normal frames.

Figure 15.3 shows an example of the startup phase of the network.

> **NOTE**
>
> If there are multiple leading nodes (which may be the case if it has been considered that one of them may
> fail), the node with the smallest number of slots wins, and it transmits its startup frame in cycle 0 and then
> becomes the leading coldstart node.

Now that the network has woken up, has started, is functioning, we need to remember
that our world is not totally marvellous and errors do happen in it, so let us now go on
to look at how they are managed!

15.3 Error Management

15.3.1 'Never Give Up' Strategy

Let us begin by explaining in a few simple words the background of the strategy for
errors that may occur. As we will show, to have the most reliable possible system,

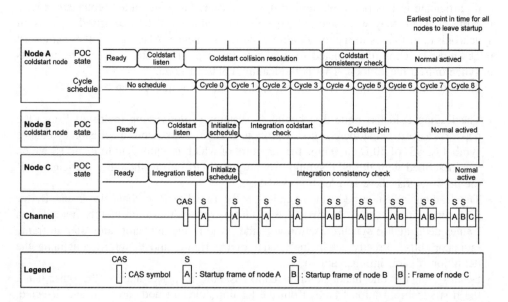

Figure 15.3

FlexRay operates as a 'never give up' device. This means that in the case of problems on the network, the whole system never throws in the towel, always tries to operate and never gives up because of an incident. To do this:

- if a communication system is unavailable, this must inhibit distributed assistance (or salvage) mechanisms;
- the act of restarting (in operation, hot) a node (which has failed for some reason) in a system implies more than just restarting the communication part of the system (for example, restoring the application context, and so on).

Why? It's very simple:

- in general, the communication system represents only the most visible part of a much larger system, the overall purpose of which is to serve specific applications;
- errors of all types can appear anywhere, throughout the system;
- some particular errors can be recognised, diagnosed and dealt with only by the application layer;
- operating modes based on processes of 'end-to-end' agreement (end-to-end systems) and interactive consistency protocols make it possible to detect communication errors on the fly.

Consequently, the approach of the solution can be summarised as:

- on the one hand, maintaining data transmission for as long as communication between the other nodes is not compromised, that is:
  - for as long as synchronisation of fault-tolerant clocks functions;
  - for as long as the information from the key operation and the transmission is that the checks and health judgements have been passed successfully.
- on the other hand, maintaining reception for as long as possible, that is:
  - here too, for as long as synchronisation of fault-tolerant clocks functions.

15.3.2   Error Management

As we have just stated, the FlexRay error management philosophy is principally demonstrated in a 'never give up' strategy and robustness against transient faults. Consequently, the model of error management and detection tries to avoid 'faulty' behaviour in the presence of faults, and must support a specific degradation concept.

This concept is closely linked to the severity of an error caused by the system, or by the repetition that the same error can cause during a certain elapsed time.

FlexRay, on the one hand, defines four levels of degradation depending on a classification of possible errors and, on the other hand, specifies the management and classification of errors that occur depending on levels/classes of severity, 'Sx':

- **Class S0 – Normal operation** – sending and receiving, full operation with the CC and line driver.

- **Class S1 – Warning** – full operation with the CC and line driver, and the host processor is warned.
- **Class S2 – Error** – transmission is stopped, and the CC and line driver remain synchronised. The host processor is warned.
- **Class S3 – Fatal error** – operations are stopped. All the pins are put into 'safe' state. The host processor is warned, and the line driver blocks access to the lines of the network.

Figure 15.4 gives a summary of the state diagram, indicating the transitions according to the external conditions.

Additionally, the error management must allow return to normal operation if the error conditions no longer exist according to their levels.

15.3.3 States of the Protocol

The error management philosophy above and its degradation model, including the severity levels 'Sx', influence the states of the general protocol. This degradation goes through the following states in succession:

- Normal
  - Normal static
  - Listen only

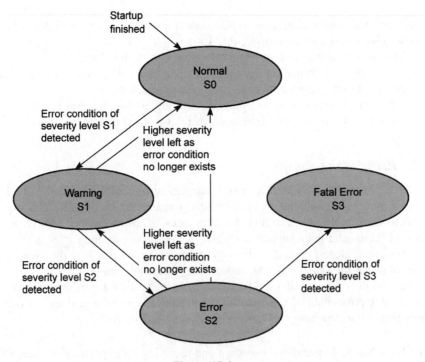

Figure 15.4

  – Normal slave
  – Normal master
- Passive
- Error.

Figure 15.5 shows these states and how the transitions from state to state take place according to the severity conditions.

15.3.4 Errors on the Channels and Communication Frames

FlexRay also defines specific error-detection mechanisms for errors that occur on the communication channels and the frames that circulate on them. Error detection concerning the channels and the associated 'host' information make it possible to manage all the traffic and the errors on the medium. Two types of information called 'channel vectors' and 'frame vectors' characterise these error states.

15.3.4.1 Channel Status Error Vector (CSEV)

The so-called channel errors (bit coding, CRC error, slot error, cycle counting error and frame length error) trigger a CSEV, and linked to the associated host information, they make it possible to have a specific observation of the frames of a certain node.

These CSEVs are located in the interface of the FlexRay CC, and can be configured to be the sources of interrupts.

The CSEVs are reset explicitly by the host.

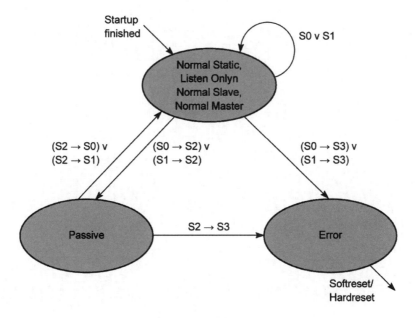

Figure 15.5

15.3.4.2   Frame Status Error Vector (FSEV)

It is the same for frame errors (bit coding, CRC error, slot error, cycle counting error, wrong frame length, missing frame and null frame), which trigger an FSEV, and linked to the associated host information, they too make it possible to have a specific observation of the frames of a certain node.

These FSEVs are located in the interface of the FlexRay CC, and can be configured to be the sources of interrupts covering all the frames in a CC.

The FSEVs are reset explicitly by the host at the end of each hoped-for frame.

This approach fulfils the 'need to know' philosophy of FlexRay.

16

FlexRay v3.0[1]

For years, FlexRay architectures have been well known for their efficiency. Now this communication protocol/standard is mature and many microcontrollers with embedded FlexRay interfaces from several semiconductor manufacturers have passed official conformance tests. FlexRay projects are running in all automotive Original Equipment Manufacturers (OEMs), and many experiences from several projects have been incorporated to create the ultimate release of the specifications. The FlexRay Consortium worked over nine years to create the final set of this FlexRay standard for In-Vehicle Automotive Networking, and 'version 3.0' of the FlexRay specifications was finally published in mid-December 2009. This included:

- FlexRay communications system – protocol specification (PS), v3.0;
- FlexRay communications system – electrical physical layer (EPL) specification, v3.0.

All the principles of 'version 2.1' of the specifications, which we have described throughout the previous chapters, remain the same, and the majority of the changes in 'version 3.0' have been introduced to guarantee full performance of a FlexRay system at the limits of operation. In addition, most of the requirements from the Japanese market[2] have been taken into account in order to create a worldwide standard.

16.1 Protocol Enhancements

In one short chapter we will not even try to summarise all the new items contained within the 336 pages of the v3.0 PS document; we will merely present a brief list of

---

[1] The protocol enhancements part of this chapter is based on published documents from Peter Spindler, Systems Engineer at FreeScale Semiconductors. For the physical layer enhancements part, I would like to thank two ex-colleagues and friends – Matthias Muth and Steffen Lorenz, both involved for many years in the physical layer and ISO process at NXP Semiconductors, AIC, Hamburg – for the information, documents and authorisation they have given me to use and publish these paragraphs.

[2] The Japan Automotive Software Platform and Architecture (JASPAR) Consortium has been continuing coordinated activities with the FlexRay Consortium with the aim of ensuring a unified international standard of the next generation in-car network. As a result, it was decided that a technical proposal from JASPAR should be adopted in the FlexRay Protocol Specification v3.0, the Physical Layer Specification v3.0 and the FlexRay Conformance Test Specifications. This will enable semiconductor manufacturers to build a single device that serves a global market.

FlexRay and its Applications: Real Time Multiplexed Network, First Edition. Dominique Paret.
© 2012 John Wiley & Sons, Ltd. Published 2012 by John Wiley & Sons, Ltd.

the main points of protocol enhancements. For full details we encourage you to read the complete document!

- **Bit rates**
  In previous versions of the specification, bit rate was only defined at 10 Mbit/s and other bit rates only appeared sporadically in some specific tables. Now FlexRay supports officially 2.5, 5 and 10 Mbit/s.
- **Slot multiplexing**
  Now, sharing of static communication slots – slot multiplexing – is possible between multiple nodes.
- **First in, first out (FIFO) buffer**
  Now, at least one FIFO buffer is mandatory and details are given for FIFO filter criteria.
- **Cycle counter**
  Concerning configurable cycle counter wraparound, now, any even number between 8 and 64 is possible and extended cycle counter filtering can take repetition values of 5, 10, 20, 40 and 50.
- **Timers**
  There are no relative timers any more, but two absolute timers.
- **Controller-host interface (CHI) commands**
  New CHI commands exist for 'save shutdown' at the end of the communication cycle.
- **Individual buffers**
  It is possible to reconfigure message buffers to some extent.
- **Network management vector**
  The network management vector is mandatory now.
- **Status data**
  The number of received startup frames is provided to the host.
- **Blind phase**
  Network activity is ignored for a configurable period of time following a transmission to minimise effects caused by echoes/ringing.
- **Dynamic segment robustness**
  Robustness and behaviour are improved in case of 'noise' on an undriven link in the dynamic segment (avoid desynchronisation of the slot counter in the cluster). For example, noise does not lead to a slot counter desynchronisation in the case of:
  - short symmetric noise;
  - short asymmetric noise;
  - noise after a frame.
  note: classification of noise depends on duration of activity. An activity shorter than 80 bits is classified as noise.
  No fatal protocol error exists any more for:
  - ongoing transmission at the end of the dynamic segment;
  - automatic termination of a frame transmission in the dynamic segment if the dynamic segment ends before the frame transmission is complete.
- **New synchronisation modes, TT-E, TT-L**
  The goal of the new synchronisation modes is to provide simple means for connecting subnetworks to a FlexRay network such that synchronisation is retained between the networks. The key aspects are:

- alignment of schedules;
- synchronisation transparent to higher layers;
- scalable (single or dual channel);
- simplicity, basic solution;
- introduced time-triggered master mode (time-triggered local (TT-L) and time-triggered external (TT-E)).

16.1.1 TT-L – Time-Triggered Local Master Synchronisation

The TT-L synchronisation method is an addition to the conventional FlexRay synchronisation that we described in the previous chapter. TT-L provides a simple means for faster cluster startup, better cluster precision and synchronisation controlled by a single node. It is applicable for single or dual-channel systems and downward compatible to v2.1.

A FlexRay TT-L cluster using this synchronisation method has only a single synchronisation-frame-transmitting CC. The TT-L coldstart node, which is dedicated for startup of the FlexRay cluster, transmits two synchronisation frames in each cycle and is transparent to non-sync nodes.

- FlexRay v2.1: at least two coldstart nodes are required.
- FlexRay v3.0: a single special coldstart node is allowed (TT-L node).

As only the sole TT-L coldstart node transmits synchronisation frames within the TT-L cluster, the rate correction cannot accumulate errors and no cluster drift damping factor is required. It is useful for small stand-alone systems and for developing subnetworks for multi-cluster systems.

16.1.2 TT-E – Time-Triggered External Synchronisation

The TT-E synchronisation method is an addition to conventional FlexRay synchronisation. TT-E provides a simple means for connecting FlexRay clusters such that all clusters are synchronised. The key aspects are:

- alignment of schedules;
- synchronisation transparent to higher level/host MCU;
- applicable to single and dual-channel systems;
- backward compatible – that is, can be used to synchronise two FlexRay v2.1 clusters.

Clusters using the TT-E synchronisation method are called TT-E clusters.

Whereas in TT-D (old, conventional version) and TT-L networks all CCs use the same synchronisation algorithm, the sync frame transmitting CCs within the TT-E cluster behave differently from the other CCs, which requires the estimation of a range of precision between the various types of CC.

To understand how this works it is necessary to introduce some new terminology in order to differentiate clearly between the different clusters and nodes:

- **time source cluster** is the FlexRay cluster from which the timing is derived;
- **time sink cluster** is the one from which the timing is determined.

Both clusters together form a time synchronised cluster pair.

The TT-E coldstart node is then called a **time gateway sink**, while the FlexRay node of the time source cluster it is connected to is called a **time gateway source**.

- a **time gateway source** may be of arbitrary type – that is, a coldstart node, sync node or neither of those;
- a **time source cluster** may be of arbitrary type – that is, a TT-D cluster, TT-L cluster or even a TT-E cluster itself.

The time gateway source and the time gateway sink are connected via a **time gateway interface** propagating information of the time gateway source towards the time gateway sink.

While for the TT-L synchronisation method, only a single TT-L coldstart node may be present in a given TT-L cluster, such a restriction is not necessary for TT-E coldstart nodes in a TT-E cluster. Each time gateway sink will be synchronised with its time gateway source, thereby with the time source cluster, and thereby implicitly with all other time gateway sinks.

Each time sink cluster may have only a single time source cluster. This implies that the time gateway sinks may only be connected to time gateway sources belonging to a single cluster. On the other hand, one time source cluster may have multiple time sink clusters attached.

Via the time gateway interface, the time gateway source provides the time gateway sink with information about its current state, position of cycle start and cycle length 1. Furthermore, all terms the time gateway source derives from its clock synchronisation algorithm are forwarded to the time gateway sink. The information about the state of the time gateway source contains, at least, whether the time gateway source is in normal active, normal passive or another state. At various places, it is necessary to refer to configuration parameters or precision estimates of the time source cluster.

So, as conclusions:

- **TT-E uses case 'external sync mode'**
  - TT-E is transparent to non-sync nodes;
  - TT-E coldstart node:
    * is used for externally synchronised networks;
    * is able to retrieve startup and synchronisation information from another cluster and provide this information to the TT-E cluster;
    * can switch autonomously to local sync mode in order to keep communication within the TT-E cluster ongoing if a TT-D cluster is not active.
- **TT-E synchronisation method**
  - single gateway node connects both FlexRay clusters;
  - gateway node: time gateway source node + time gateway sink node;
  - time gateway interface: clock synchronisation;
  - time gateway sink node: TT-L node – that is, single node which drives synchronisation in sink cluster TT;
  - as usual, same cycle length in both clusters;
  - fixed cycle offset between clusters;

– clusters can have different schedules (that is, different numbers of static slots, sizes of static slots, lengths of dynamic segment, lengths of minislots or lengths of network idle time (NIT)).

16.2 Physical Layer Enhancements

Revisions A and B of version 2.1 showed some lack of specification and room for improvement. Version 3.0 is an unambiguous description and natural progression of the physical layer and closes gaps in previous versions. It gives additional measures for signal integrity (SI) to give a very good definition of reliable network topologies.

The interoperability with former specifications is kept throughout version 3.0 of the EPL specification and even though many of the new features and more specific timing requirements are already provided by some EPL 2.1 compliant products, the EPL 3.0 standard ensures that these will be in every product in the future. The electrical physical layer application notes (EPLAN) of the EPL 3.0 have been expanded to give valuable information for the implementation of FlexRay systems.

The major advantages of EPL 3.0 and EPLAN 3.0 compared to the previous version are:

- SI voting and changed eye diagrams provide an unambiguous assessment of SI and thus reliable topologies;
- the more restrictive specification of asymmetric delay-related parameters avoids unnecessary limitations of possible topologies;
- tightened parameters perfect the system behaviour, for example faster exclusion of babbling idiots;
- newly introduced parameters (e.g. idle loop delay needed to rely on system behaviour, for example guarantee a proper startup even at the limits of operation);
- wakeup via frames allows a bandwidth-optimised wakeup during operation;
- full description of the active star guarantees reliable interaction with FlexRay CCs, for example timeouts and error confinement;
- compatibility with former FlexRay EPL versions allows heterogeneous networks, for example interoperability of new electronic control units (ECUs) with already existing ECUs;
- alignment with JASPAR requirements provides a worldwide standard and a strong basis for becoming an ISO standard.

The following paragraphs give the main technical changes, the motivation for changing and the consequences of such changes as well as the interoperability between the different versions.

16.2.1 From Network Implementation to Signal Integrity Focus

As previously described, the intention of EPL 2.1 was to have all topology parameters (cable attenuation, cable length and number of nodes) limited in order to allow an easy network implementation. Though the probability is high of finding a suitable combination, it has been proven that it cannot be guaranteed that all possible combinations of the specified parameters provide a reliable network topology. Therefore, the original approach was changed. The EPL 3.0 focus is now set to SI, guaranteeing appropriate decoding at receivers and unnecessary limitations are removed, thus allowing more flexibility.

16.2.2 Signal Integrity Improvements

In this new approach, measures are needed for SI with a clear distinction between contribution of the device and contribution of the network. EPL 3.0 includes several timing and voltage level requirements for the assessment of transmitter and receiver of line driver devices, guaranteeing a minimum output signal. To provide additional physical measures for the devices, mask tests have been introduced to summarise the transmitter output requirements.

New methods are now included in EPLAN 3.0 for the quality of the network and for the validation of different topologies.

The eye diagrams (see Figure 16.1) previously included in the EPL have been shifted to EPLAN. The definition of the eye diagrams and the method of capture have been adapted in order to include the requirements of time-triggered behaviour. With this definition of the eye diagrams, the consideration of the minimum bit duration is implicitly included. For the different data rates, dedicated eye diagrams are now available to consider the resulting differences of the decoder's minimum timing requirements.

EPL has also defined two receiver thresholds, distinguishing three states on the bus wires: Data_0, Data_1 and Idle. This provides a certain robustness against ringing effects, which are not completely considered in the eye diagram. A differential signal that violates the eye diagram must not necessarily cause a decoding error. To overcome this over-strictness, SI voting (a mathematical procedure) has been introduced. SI voting finally judges whether the resulting RxD signal shape allows fault-free decoding. The SI voting

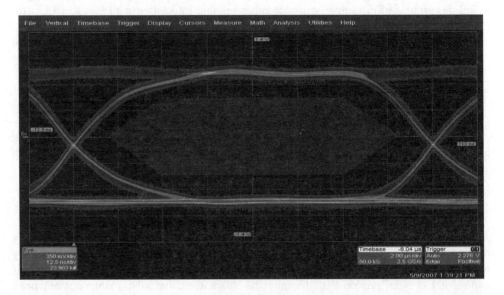

Figure 16.1

procedure is implemented as a function in specific oscilloscopes (for example the LeCroy oscilloscope – see Figure 16.2).

A failure in the eye diagram test in SI voting indicates a problem in the network, which concerns the topology and network parts (cables, connectors, chokes, PCB, and so on).

Beside these two physical measures, as usual the simulation of the network behaviour is highly recommended as an additional tool for defining the right topology.

16.2.3   Timing Improvements

The FlexRay protocol has certain timing requirements which have to be fulfilled by the physical layer. As shown in detail in another chapter, most critical is the asymmetric delay budget of the signal path from the signal source (protocol engine in the transmitting CC) via the network to the signal sink (protocol engine in the receiving CC) to guarantee a proper decoding of the data stream. The asymmetric delay budget describes the maximum allowable bit deformation (lengthening or shortening) that can be accepted by the decoder. The bit deformation is caused by different propagation delays of the rising and falling edges.

With EPL 2.1A, a worst-case calculation of the path from transmitting BD to receiving BD was already possible, but due to the lack of stringent requirements, the results were

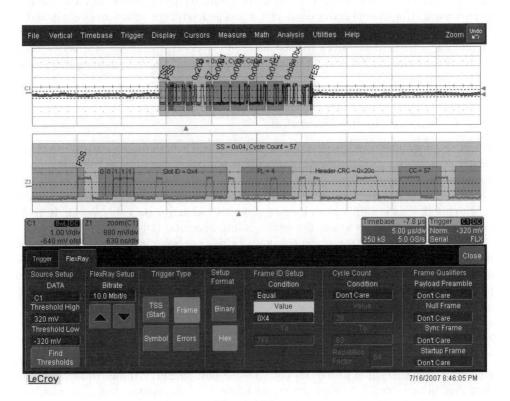

Figure 16.2

often too pessimistic and associated with unnecessary limitations for the topologies. In EPL 3.0 there is a description of the complete signal path consequently realised, with improvements in several network parts. The signal path diagram now includes requirements for all components between the transmitting and the receiving protocol engine, for example the CC and the BD-CC interface. The BD-CC interface (TxEN, TxD and RxD) is clearly specified with thresholds, output levels, rise and fall times and load conditions.

In addition, the interactions with the FlexRay PS and the general system performance were improved by several changes of existing system parameters and by introducing new parameters. For example, the decreased limit of the TxEN timeout is limiting the effect of erroneously permanent transmitting nodes ('babbling idiots') and the newly introduced 'idle loop delay' is needed to guarantee a proper startup of the system.

Another important point: almost all FlexRay topologies currently implemented in cars use active stars, which allow a high number of nodes to be connected in a network by keeping the SI at a high level. The error confinement capabilities are able to isolate erroneous branches while continuing communication on the rest of the network. As a consequence of its functionality, the active star is a central element in the network and therefore reliable behaviour is needed. With EPL 3.0 the active star chapter has been completely reworked, resulting in a comprehensive and unambiguous specification.

Additional timing parameters have been introduced and correlations exploited to reduce the range of existing timing parameters. The optimised timing parameters facilitate the protocol constraints, allowing a more efficient parameterisation of the protocol. Several improvements (for example error confinement and undervoltage behaviour) were included to optimise and standardise the functional behaviour, and other timing parameters were introduced to have fixed values for standardised drivers (for example automotive open system architecture (AUTOSAR)); mode transition time and undervoltage recovery time are just two examples.

16.2.4 Wakeup During Operation

In EPL 3.0, for all bit rates, the description of the wakeup pattern for remote wakeup of bus drivers or active stars via the FlexRay network is now more precisely and unambiguously documented.

The normal wakeup pattern is sent during the startup procedure before the network is synchronised. This is sufficient as long as all nodes and active stars stay awake. In some cases, the nodes shall be woken up, again, during operation (after the synchronisation). One solution is to send the wakeup pattern in the symbol window but, to overcome the problem of using a lot of bandwidth on symbol windows, the EPL 3.0 describes a dedicated payload for wakeup via frames which can be transmitted as a normal payload in any data frame.

16.2.5 Interoperability of Different EPL Versions

The key factor for the interoperability of different physical layer devices in one FlexRay network is the interaction of the devices with the bus. In EPL 3.0 the requirements are described in more detail (voltage amplitude transmitter, transmitter slopes, mismatch of the slopes, ringing effects, electromagnetic compatibility (EMC) behaviour, and so on) whereas the behaviour remains unchanged. However, additional new limits improve the results of worst-case calculations of the asymmetric delay from sending to receiving CC, and topologies with long distances or many stubs will benefit from it.

As the receiver is unchanged, devices that do and do not implement functional class can operate concurrently in the same network. The limiting factor is the SI, which can be assessed with the eye diagrams or SI voting, as described above.

As a global conclusion, the different EPL versions are interoperable.

- EPL 3.0 is designed to make maximum use of the PS 3.0. It is the basis for protocol constraints and worst-case calculations.
- EPL 3.0 compatibility with a PS 2.1-compliant CC is still given.
- A bus driver compliant with EPL 3.0 automatically exceeds the EPL 2.1 requirements.
- Concerning the reusability of existing ECUs:
  - the configuration of a CC compliant with PS 3.0 allows operation in heterogeneous networks with CCs compliant with PS 2.1;
  - however, the full enhanced feature set of PS 3.0 will only be available in homogeneous PS 3.0 networks.

The different combinations of BD and CC are summarised in Table 16.1.

As a short final conclusion, for new active star applications and heterogeneous networks it is highly recommended to use 3.0-compliant devices, as only with EPL 3.0 can the required active star behaviour and timing be guaranteed. Additionally, again, it is recommended to assess the topologies by simulation.

16.3   FlexRay and ISO

By October 2010, with the very last, 'v3.0.1' package (see Figure 16.3) of the specifications, the official FlexRay Consortium's life was over.

Table 16.1   Interoperability of different BD and CC versions

BD	CC	Comment on interoperability
2.1	2.1	Bus driver does not automatically provide all necessary parameters and features BD-CC interface not sufficiently defined (timing calculation to be performed individually)
2.1	3.0	Bus driver does not automatically provide all necessary parameters and features BD-CC interface not sufficiently defined (timing calculation to be performed individually) Some 3.0 protocol features are not applicable
3.0	2.1	Bus driver provides all necessary parameters and features CC part of the BD-CC interface not defined (timing calculation of EPLAN 3.0 to be modified to actual CC values)
3.0	3.0	Bus driver and BD-CC interface unambiguously defined, timing calculation given in EPLAN 3.0

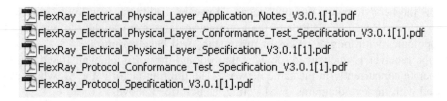

Figure 16.3

All of the specifications have already reached a high technical and redaction quality level and, as a consequence and in the normal manner, these FlexRay specifications have been transferred to the appropriate committee of the International Organization for Standardization (ISO), *ISO/TC 22/SC 3/WG 1-Road vehicles*, as a 'New Work Item Proposal (NWIP)'. The approval and release of the work as an internationally approved 'IS' standard will take place over the next one to two years and it is expected to have no functional or technical changes within the ISO process.

NWIP 17458-1 – FlexRay communications system – Part 1: General information and use case definition – N 3065	2011-08-19
NWIP 17458-2 – FlexRay communications system – Part 2: Data link layer specification	2011-08-19
NWIP 17458-3 – FlexRay communications system – Part 3: Data link layer conformance test specification	2011-08-19
NWIP 17458-4 – FlexRay communications system – Part 4: Electrical physical layer specification	2011-08-19
NWIP 17458-5 – FlexRay communications system – Part 5: Electrical physical layer conformance test specification	2011-08-19

As a final point, of course, new certifications will come to be conformant to ISO tests!

16.4   FlexRay in Other Industries

From its earliest beginnings, the FlexRay Consortium, represented by its core partners, had the goal of defining a new communication standard for the automotive industry without any undue influence from outside parties. As an example, some lines from the 'Version 3.0 Disclaimer':

'This specification and the material contained in it, as released by the FlexRay Consortium, is for the purpose of information only. The FlexRay Consortium and the companies that have contributed to it shall not be liable for any use of the specification. ... The word FlexRay and the FlexRay logo are registered trademarks.'

but also, in the 'Version 3.0 and 3.0.1 Disclaimers':

> '... The FlexRay specifications have been developed for automotive applications only. They have neither been developed nor tested for non-automotive applications.'

Coming back to this text, '*developed ... only*' does not imply 'reserved for' or 'reserved strictly for.' It is to say '*They have neither been developed nor tested for non-automotive applications*' but it is not explicitly forbidden ... in others words there is some authorisation for other uses and applications (industrial, aeroplanes, and so on).

To date – with or without ISO publication – there is some visibility of FlexRay-based applications outside of the automotive domain. Some (small) confusion is still caused by the debate concerning explicit licences to the essential intellectual property rights (IPR) for 'non-automotive' applications, which would facilitate specific non-automotive FlexRay products. Fortunately, users are content to use FlexRay products designed for automotive applications.

It is important to note that companies wishing to utilise products certified as being compliant with FlexRay do not need to acquire their own licence to the essential IPR, as long as the products have been brought onto the market under a licence in accordance with the FlexRay agreements (like for CAN). Since the FlexRay devices currently on offer by all semiconductor vendors meet this requirement, users of such devices are free to use these devices however they wish to, without having to obtain further permission from the core partners, which is also in line with the legal concept of 'exhaustion'.

Part E

Architecture of a Node, Components and Development Aid Tools

17

Architecture of a FlexRay Node

17.1 The Major Components of a Node

The conventional architecture of a FlexRay node consists of four major blocks. Three of them are necessary and the fourth can be considered as optional. They are:

- the 'host' microcontroller, which holds the application and the application management of the communication protocol;
- the protocol manager, which constructs the cycles, segments, and so on, the famous communication controller 'CC' which has already been mentioned several times;
- the line driver, also described in Chapter 8;
- and finally, the 'bus guardian'.

Figure 17.1 details the contents of the internal/external connections of a FlexRay node between the host controller, which manages the application overall, the FlexRay protocol manager, the line driver(s) (the 's' in brackets being associated with applications with a single or double communication channel), and finally, if necessary, the bus guardian.

It should be noted that, very often, the protocol manager and the host microcontroller form one and the same integrated circuit, and that today there are few solutions which use a bus guardian. Also, certain technologies which exist on the market (power current, ESD protection, and so on) lend themselves well to the future existence of line drivers and bus guardian integrated on the same silicon.

Figure 17.2 shows more precisely the logical relationships that exist between the different parts of the block diagram above. This figure also shows clearly (when it is read from bottom to top) how the architecture of a FlexRay node conforms to the separation of layers 1 and 2 on the one hand and 7 on the other hand of ISO's OSI model. Conforming to this separation enables everyone (vehicle manufacturers, equipment manufacturers, and so on) to develop, easily and independently, each of the entities shown in the figure.

17.2 Architecture of the Processor and Protocol Manager

The specifications dedicated to the FlexRay protocol describe in great detail the architectures that the interfaces of the host processor and protocol manager must have (block diagrams, software routines and subroutines – which are practically mandatory for any

FlexRay and its Applications: Real Time Multiplexed Network, First Edition. Dominique Paret.
© 2012 John Wiley & Sons, Ltd. Published 2012 by John Wiley & Sons, Ltd.

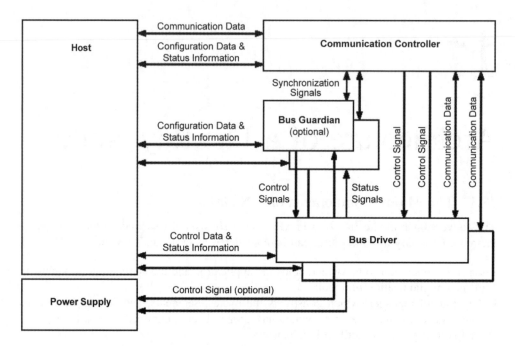

Figure 17.1

hope of passing the conformity tests), but also the necessary architecture that a node of the network must have to ensure that the network functions well. These descriptions are very oriented to semiconductor manufacturers, integrated circuit designers, component designers – controller either separate or integrated on the microcontrollers – whether they are in FPGA or ASIC technologies, OEM solutions, and so on. Despite 40 years spent with one of the biggest leaders in integrated circuits in this field (Philips Semiconductors/ NXP, but we won't name it!), we won't condemn you to the torture of dissecting, with us, all these documents to arrive at the level of NAND gates. In this chapter, we will merely skim through these long specifications in a few words.

Before we begin, let us just make a small, but important, final, basic comment. Like any protocol specification, this one describes what must be satisfied (to satisfy the conformance tests), but definitely not how it should be implemented in silicon. Everyone will have their own tricks, clever ways of doing things, and so on. There will therefore be, as always, different hardware and pseudo software implementations, such that (i) they conform, (ii) they are interoperable and (iii) the consistency of the network is fully respected. After everything we have written on these subjects, that should seem clear and obvious to you . . . but!

Figure 17.3 shows the internal architecture of the protocol controller as recommended by the FlexRay specification, and you will recognise all the functional modules which we have described in great detail in the earlier chapters (generation of MTs, synchronisation, and so on). To finish on this subject, before we have actually said anything, it is normal

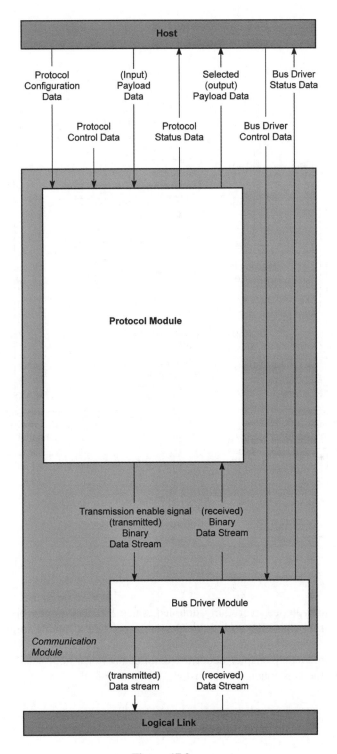

Figure 17.2

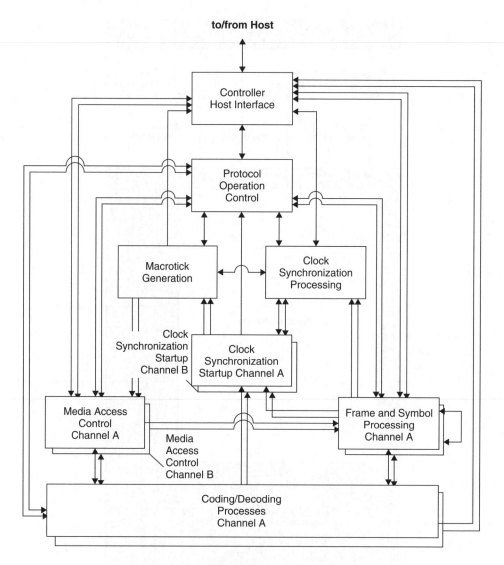

to/from Host

Figure 17.3

that in this figure you see everything doubled, since FlexRay is capable of supporting two communication channels, so it is necessary to provide a double implementation of the functions.

This short introductory chapter about the architectures of a node leads us naturally to be interested in the components that exist to satisfy them!

18

Electronic Components for the FlexRay Network

Before going into mass production, advanced systems and new technologies can only be introduced to the market via the arrival of new generations of top of the range car models, which are expensive, as happened with CAN when it began. Now, the planet does not hold many such models, and they don't reproduce like rabbits! Also, the whole of industry will tell you that starting from scratch, to design a new architecture of aircraft takes about 30 years, a high-speed train 15–20 years and a car about 7 years. Once this rule is known, it is easy for a manufacturer to determine, without going too far wrong, what will be the next date when a new model or new platform will appear.

All the crystal balls got it wrong. For several years, it was known that the most probable window for launching such a project industrially was late 2006 or early 2007. So if you count backwards a little, you will not be surprised to know that everyone was hard at work from the years 1999/2000, and that three years before the launch (so in 2003), most of the electronic components were already in the phases of preliminary or final certification. Surprising, isn't it? In short, the silicon wafers of the line drivers (from Philips/NXP Semiconductors) and communication controllers (from Freescale Semiconductors, formerly Motorola Semiconductors) have already been leaving the diffusion ovens for some time – at the same time as the supply forecasts from the vehicle manufacturers were being refined.

For historical interest, Figure 18.1 shows the initial plan for introducing the FlexRay project to the market – which in fact has been practically entirely followed! When that happens for once, it deserves to be emphasised!

18.1 The Component Range

Getting back to basics, you should now know that the FlexRay set of components has a family resemblance to that of CAN. That is, it consists of protocol managers, microcontrollers including protocol managers, one or two line drivers and new members: depending on whether the applications are redundant or not, active stars and, finally, depending on

---

FlexRay and its Applications: Real Time Multiplexed Network, First Edition. Dominique Paret.
© 2012 John Wiley & Sons, Ltd. Published 2012 by John Wiley & Sons, Ltd.

FlexRay Roadmap

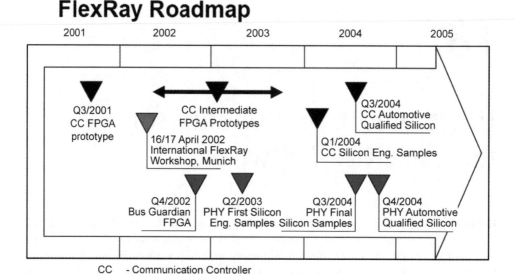

CC   - Communication Controller
PHY - Bis Driver / Bus Guardian / Active Star

Figure 18.1

the applications, a special element, a bus guardian. After the two first pioneers – Freescale and NXP Semiconductors – numerous other semiconductor manufacturers (AMS, Elmos, ST, Infineon, Fujitsu, Renesa, Samsung, and so on) also expect to be able to supply this nascent market.

Let us now make a list of the first members of this family.

18.1.1 FlexRay Protocol Manager

Freescale having participated in the design, development and evolution of FlexRay within the Consortium, it proposed the first protocol managers in the form of separate boxes, implemented in the form of programmable circuits of FPGA type. Their internal designs obviously satisfy the required architecture of the 'protocol engine', as described in the preceding paragraphs. Obviously, time passed, and the FlexRay protocol manager was quickly integrated with the host (micro)controller – for example, depending on the manufacturer, of PC Core, ARM 9, ARM 11 or H8 types – which also manages the application.

A generic example of a block diagram of a microcontroller for an integrated FlexRay manager, which also provides the link to the CAN and LIN protocols, is shown in Figure 18.2.

18.1.1.1 Protocol Controller

With good intentions and without excessive publicity, Figure 18.3 shows the first family of Freescale microcontrollers with the FlexRay manager modules on board them.

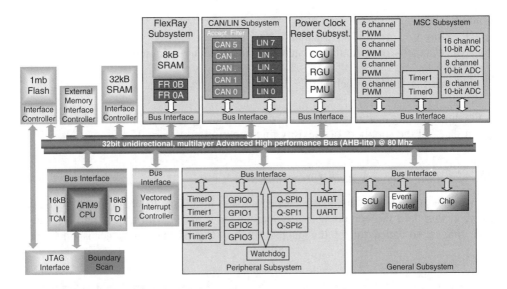

Figure 18.2

Applications	Design Considerations	Product
X-by-Wire Main Controller	Computing power Advanced safety features	MPC5561, MPC5567, MPC551xG
X-by-Wire Satellite Nodes	Small footprint Low cost Advanced safety features	S12XF-family, MPC560xP
Power train	Computing power Advanced timing features (eTPU)	MPC5567, MPC5674F
Gateway	CAN, LIN, FlexRay communications Low power	MPC551x

Figure 18.3

Figure 18.4 shows the block diagram of the MPC 5510.

As another example, around a 16-bit HCS12X 40 MHz core to which a co-processor in XGATE is added, the MC9S12XFR comprises:

- on the one hand:
  - 128 kB of flash memory with error correction code (ECC),
  - 2 kB of programmable read-only memory – EEPROM,
  - 16 kB of RAM,
  - 16 analogue to digital conversion (ADC) channels (resolution 8 or 10 bits),
  - six PWM channels;
- on the other hand, an embedded CAN 2.0 A/B controller – (MSCAN);

- and finally:
  - FlexRay v2.1 module – 10 Mbit/s – two channels,
  - two channels for redundant systems or independent operations to double the bandwidth,
  - 32 message buffers, each with a depth of 254 bytes of data.

Obviously, other manufacturers have followed closely behind. Figure 18.5 shows, on a given date, all the suppliers of FlexRay microcontrollers.

18.1.2 Line Drivers and Active Stars

As we indicated at the start of the chapter, it was necessary to begin at the beginning. Now, in this field the beginning is the market for vehicles with a high rate of innovation; that is, so-called top of the range vehicles. Top of the range means high prices

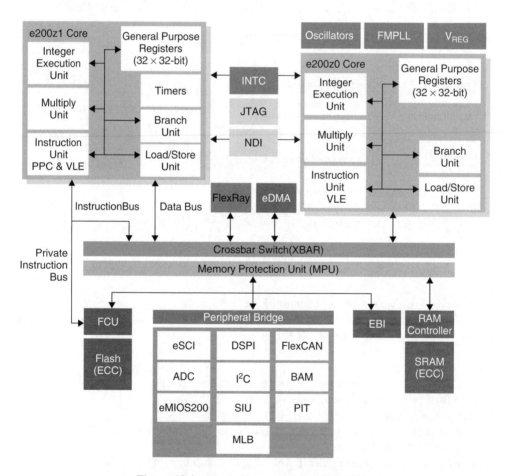

Figure 18.4 Block diagram of the MPC 5510

Mikrocontroller	Busbreite	Prozessortakt	Flash	SRAM	CAN-Schnittstellen	Besonderheiten	Anwendungsbereich
Freescale MPC5567	32 bit	40 bis 132 MHz	2 Mbyte	80 Kbyte	5	eTPU, Ethernet	Powertrain, Gateways
Freescale MPC5561	32 bit	40 bis 132 MHz	1 Mbyte	224 Kbyte	2	-	Fahrerassistenzsysteme
Freescale MPC5516G	32 bit	40 bis 80 MHz	1 Mbyte	64 Kbyte	6	MPU, eMIOS	Low-Power-Anwendungen, Gateways
Freescale MC9S12XF	16 bit	50 MHz	128 bis 512 Kbyte	16 bis 32 Kbyte	1 bis ?	XGATE, internes EEPROM	Komfort-, Chassis-, Sicherheitselektronik
Fujitsu MB91F465XA	32 bit	100 MHz	544 Kbyte	32 Kbyte	2	?	Fahrerassistenzsysteme
NEC V850E/PH03	32 bit	128 MHz	1 Mbyte	60 Kbyte	2	extra 32 Kbyte Data-Flash	Chassis- und Sicherheitselektronik
NEC V850E/CAG-4M	32 bit	80 MHz	512 Mbyte	60 Kbyte	6	MediaLB-Interface, extra 32 Kbyte Data-Flash	Komfort- und Chassiselektronik
NEC V850E/PJ3	32 bit	64 bis 128 MHz	256 bis 768 Mbyte	?	1 bis 2	?	Komfort- und Chassiselektronik
Renesas R32C/100F	32 bit	80 MHz	256 bis 512 Mbyte	32 Kbyte	2	extra 4 Kbyte Data-Flash mit EEPROM-Emulation	Komfort- und Chassiselektronik
Renesas SH725x	32 bit	200 MHz	1 bis 4 Mbyte	48 bis 256 Kbyte	1 bis ?	?	Powertrain

Figure 18.5 Example of existing components on the market

and low produced quantities, and therefore a low return on investment relative to long preliminary development. It was in this environment that the first integrated circuits were designed – obviously knowing well that a different philosophy would have to be applied later, as we will see.

Therefore, not knowing precisely the form that the architectures and final topologies of the first manufactured vehicles would take, versatile integrated circuits for multiple applications had to be implemented, so as to be able to satisfy all the first applications and to increase artificially the quantities of integrated circuits that were produced, to minimise their costs to the maximum extent when they were first introduced. This was what was done, for example by Philips/NXP SC, on request from the FlexRay Consortium, so that the TJA 1080 line driver would be multipurpose and capable of fulfilling both the simple function of a line driver and that of active stars.

Obviously, once a top of the range vehicle has been developed, everyone aims to go down into models of the less expensive range and products in larger volumes. Also obviously, costs are reexamined even more carefully than before – and revised downward! (it's strange but it's never upward, work that out!) And obviously, all that is done by thinking harder and optimising functions (simpler nodes, special stars, and so on).

After this little exercise in technical applied philosophy, let us now look at some families of line drivers.

18.1.2.1 Simple Line Drivers

To stay simple, let us say that in their broad outlines, the FlexRay drivers were designed in the same spirit as the CAN drivers. Then some special features were added because of the FlexRay protocol.

TJA 1080

For the same reasons, of belonging to the core members of the FlexRay Consortium and having participated very actively in the development of the physical layer, the first line driver to be introduced to the market was the TJA 1080 from Philips/NXP Semiconductors – obviously meeting the v2.1 physical layer specifications from the FlexRay Consortium, and having successfully passed the famous 'conformance tests' (see below).

Its block diagram is shown in Figure 18.6. There is nothing special to point out, apart from two particular interfaces for the use of the bus guardian and the host, which of course do not exist in the case of CAN.

TJA 1082

This circuit (see Figure 18.7) which is more recent than the previous one, is intended to be compatible with FlexRay v3.0, and is capable of managing bit times of 60 ns instead of 100 ns (i.e. badly damaged by asymmetrical delays). It is also capable of functioning at 42 V, which is a must for electric vehicles which have numerous power components to control and require this power supply voltage (for example, returning to the example which we described in Chapter 9, to control the motors of the worm screws of the brakes).

It is also capable of detecting bus errors and having completely passive behaviour (no untimely action) on the network when it is not powered.

Finally, to indicate its timing performance, its asymmetrical delays are 3 ns for Tx and 4/5 ns for Rx.

Since FlexRay arrived on the ground, other line driver integrated circuits with similar functions have appeared on the market. As a reminder, let us cite the best known of them:

Elmos	E 910-54, E 910-55 and E 910-56
Austria MicroSystems	AS 8220 and 8221

Today, as for CAN line drivers, the principal differences between all these products generally concern power supply options, wakeup functions by the power supply, by resets, by the network, by management of partial networks and the performance of asymmetrical propagation delays, to offer more functional and topological flexibility to users. In short, when this stage is reached, it's because the market is fully launched!

18.1.2.2 Multiple Line Drivers

Since the FlexRay protocol requires that the possibility of using two communication channels should be supported, making it possible to meet the requirements of applications of 'by wire' type, new members will be added to the line driver component families. To begin with, the first ones that are predicted will integrate two line drivers in the same box. Similarly, system basic chips (SBCs) (Paret, 2007) including voltage regulators, watchdogs, double bus guardians and line drivers are predicted.

Wait and see!

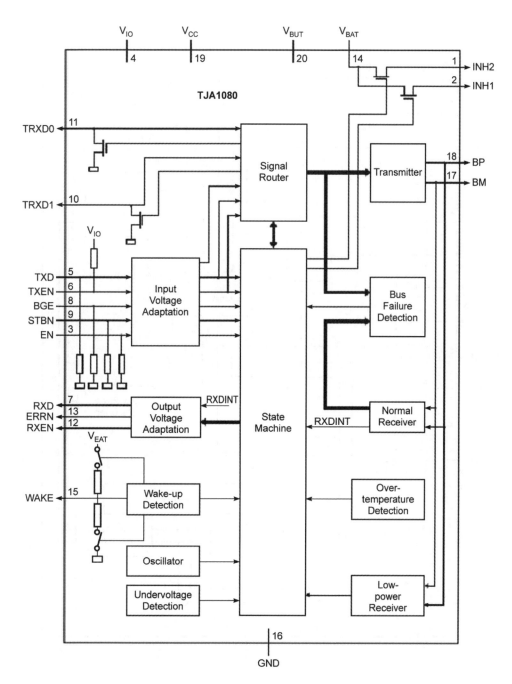

Figure 18.6

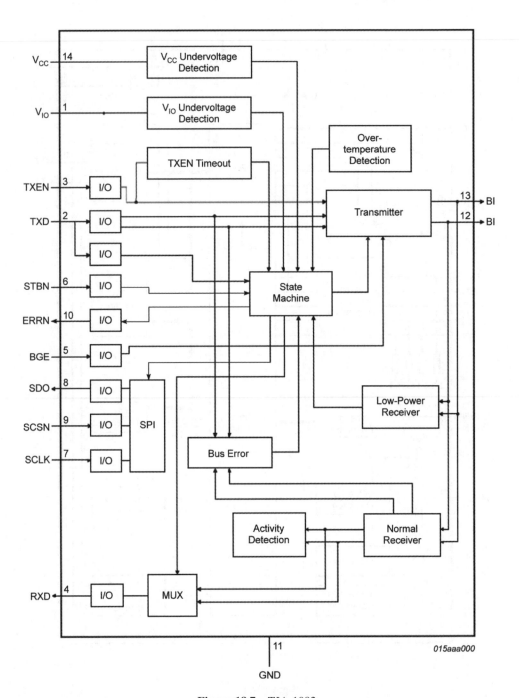

Figure 18.7   TJA 1082

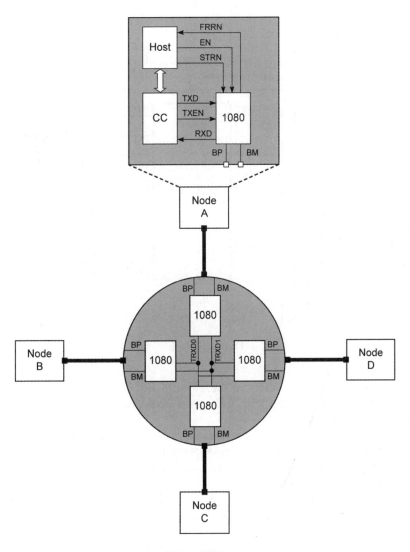

Figure 18.8

18.1.2.3 Active Stars Based on Simple Line Drivers

It is not very complicated to implement a multidirectional star using simple line drivers. The schematic block diagram is shown in Figure 18.8.

18.1.2.4 TJA 1080 in Cascade

It should also be noted that since the TJA 1080 circuit is the first line driver circuit and the FlexRay market is only beginning, as we indicated above, in order to reduce

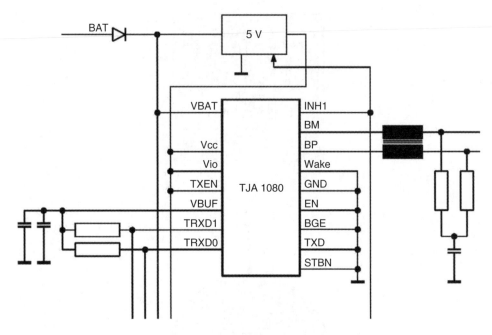

Figure 18.9

the costs of developing the circuits of the physical layer, this circuit was designed so that it could be 'cascaded', and thus active stars could easily be implemented with no additional component.

Figure 18.9 shows the schematic block diagram of this cascaded implementation, and Figure 18.10 shows its practical realisation using the TJA 1080. To make you smile a little, the photograph in Figure 18.11 shows, as an example, not merely a star but at worst the Milky Way, at best a complete galaxy!

Of course, the TJA 1080 (Figure 18.11) was designed to fulfil simultaneously the functions of a simple line driver and a cascadable driver, in order to implement active stars, time was advancing and the market was emerging, so it became necessary to optimise the costs, function by function. This is what is implemented by the new, simple line driver circuits and the specialised circuits for designing active stars.

18.1.2.5 Integrated Active Star

Let us begin with a few comments that we have (very hypocritically and very rudely) passed over in silence in the lines above about networks using active stars.

Apart from the fact that for a FlexRay network the maximum number of active stars is specified as two units, some particular comments must be made. In fact, certain applications which want to use FlexRay in generic networks, with variable geometry, have active stars from the start, but sometimes leave one, two or more branches not physically

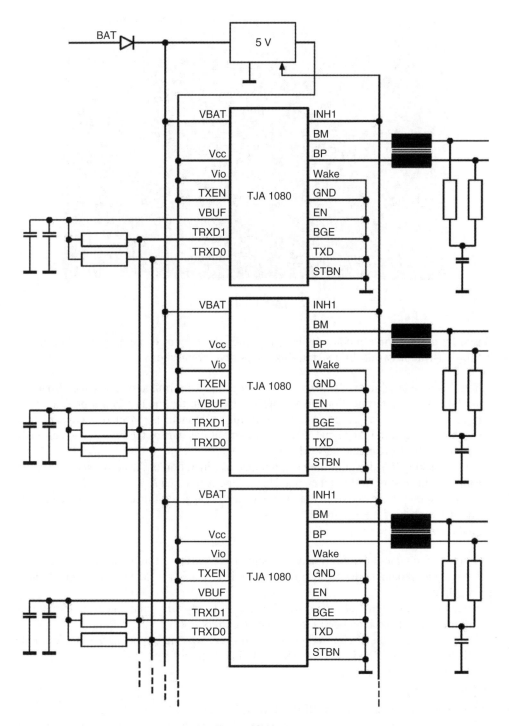

Figure 18.10

Figure 18.11

closed by an electronic module, which, for example, is considered as a possible option for forming the network. In general, this raises several questions:

- Is it the module which isn't there that has on board the termination resistance (load) of the line, or is the termination always present although the module is not present?
- If the branch is open at its end, is there, in the branch, a reflected wave which is capable of degrading the signal which is present at the start of the branch and seriously degrading the BER of the network?
- Is the attacking line driver of the branch under consideration of the active star 'blocking/isolating' in relation to the other branches of the active star?
- If not, is there, in the active star, a device which is capable of disconnecting electrically this star branch of the network?
- And so on.

Since in modular, flexible systems – including FlexRay – this type of application is not rare, you should know from now on that it is necessary to take account of these eventualities.

After these few flights of fancy, which are very heavy with consequences, let us look at the response of the integrated circuit manufacturers.

The first integrated star circuits – mono chip – are arriving on the market: for example the ELMOS 910-56 circuit and the AS 8224 circuit from AMS.

ELMOS 910-56
The very explicit block diagram of this circuit is shown in Figure 18.12.

As indicated in Figure 18.13, it can be considered either as a multiple line driver or as an active star which can drive up to four branches.

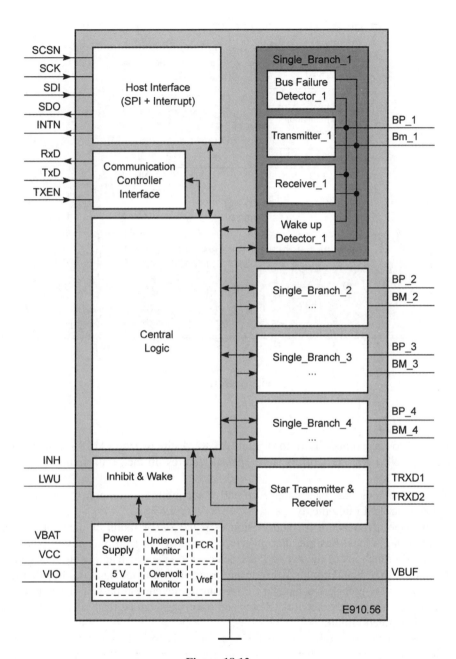

Figure 18.12

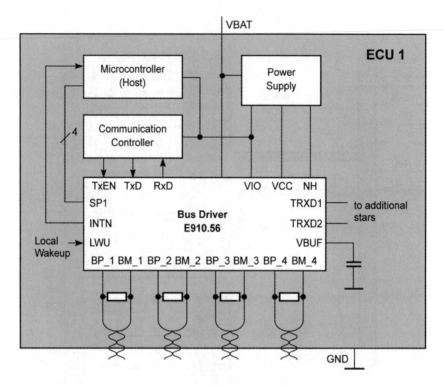

Figure 18.13

AMS – AS 8224

To respond to the points raised above, the AS 8224 circuit from AMS – responding to specification v2.1 Rev. B – offers some special features. In fact, this circuit:

- Like the previous one, can manage an active star with four branches, each of which can be controlled individually.
- Can manage a procedure for supervision of detection of faults or incidents on each branch, using a local bus guardian interface so that it can stop the activity of a faulty branch. Malfunctions of the network are detected, on the one hand, using a mechanism of analogue and digital comparisons during transmission mode, and, on the other hand, by another, very precise mechanism for measuring the current passing through the pins linking the circuit to the network. Status and error flags can be read via a host interface on the integrated circuit.
- Includes a device for reformatting bits – a 'bit reshaper'. This function is an option which is effective only when an external clock is applied to the circuit. If not, this function is bypassed and the component acts like an ordinary active star.
- Acts like an active hub to reduce the asymmetrical delay of the signal, whether it is introduced by the topology of the network or due to the components and other elements on the medium. This device is capable of reformatting single bits by steps of 12.5 ns (three μTs) up to a total of 37.5 ns, to either lengthen or shorten them. Additionally,

this mechanism compensates for drifts of the clock between the input and output flows of the circuit. The BSS of the FlexRay frame is shortened or lengthened by one µT.

- Includes an interface called 'Interstar', to enable the user to connect two or more stars in cascade, and to see only the timing performance of the whole, as if only one was present on the network.
- In total, this circuit provides six means of communication, four for the communication branches of the FlexRay network, one for managing the Interstar interface and a final one for the communication interface to the host, to provide status reporting, error signalling and other things.

18.2 EMC and EMC Measurements

Of course, since FlexRay functions at high speeds, as always, protection from disturbance of EMC type is necessary. Regarding the solution to these problems and those of radio frequency interference, same problems, same remedies! (see Figure 18.14).

In the same way as in CAN, it is recommended that terminations (impedance matching) with a midpoint should be arranged, rather than terminations without a midpoint, and common mode coils should be added if necessary. A particular comment should be emphasised in FlexRay in relation to CAN applications. In fact, they must be implemented as symmetrically as possible, to minimise as far as possible their stray inductances and to reduce the effects of asymmetrical delays as described at length in Chapter 9. For information, the annexes of the specification of the physical layer of FlexRay, on this

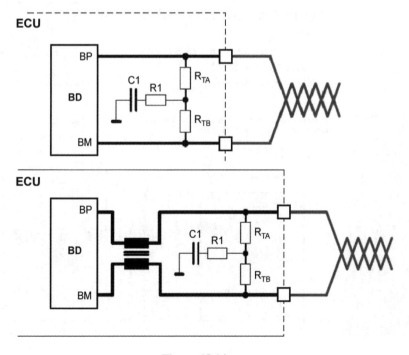

Figure 18.14

subject, indicate a maximum stray inductance value of 200 nH. This value can be reached using a coil which is implemented according to windings that have long been known, called 'bifilar windings'.

Figure 18.15 shows how this is assembled. Figure 18.16 gives the comparative results.

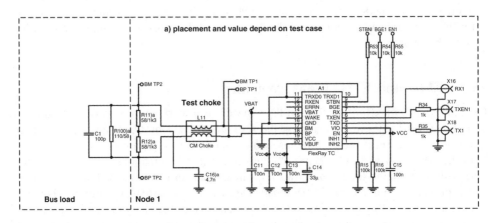

Figure 18.15

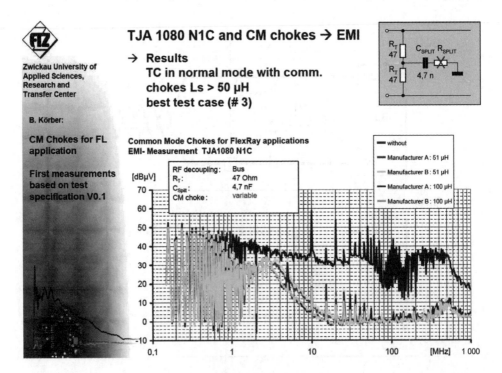

Figure 18.16

As indications on this subject, it can be said that when EMC measurements are carried out in a so-called 'direct power injection (DPI)' attack, the following results are generally obtained:

Without shared load (without 'split')	Bad
No split termination, with a choke inductance of 100 μH, low stray inductance, no capacitance	Better
Choke and termination load split	Perfect

18.3 Protection from ESD

As usual, all the electronic modules must be protected from possible electrostatic discharges (ESDs). This requirement is usually expressed by keeping to discharges of 8 kV according to the 'human model' implemented in conformity with ISO standard IEC61000-4-2, and the measurement methods recommended by the University of Zwickau.

As Figure 18.17 shows, the same protective diodes as for CAN (with low interfering capacitances), of PESD CAN diode type, can be used for FlexRay.

Figure 18.18 gives the whole conventional structure of the external components to be arranged on the pins of a line driver.

18.4 Conformity Tests

The last section of this chapter concerns conformity tests and test procedures, which the components and networks must undergo both at the level of protocol management in the strict sense, and at that of the electrical and timing operation of the physical layer.

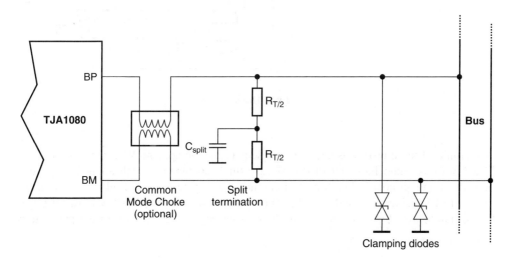

Figure 18.17

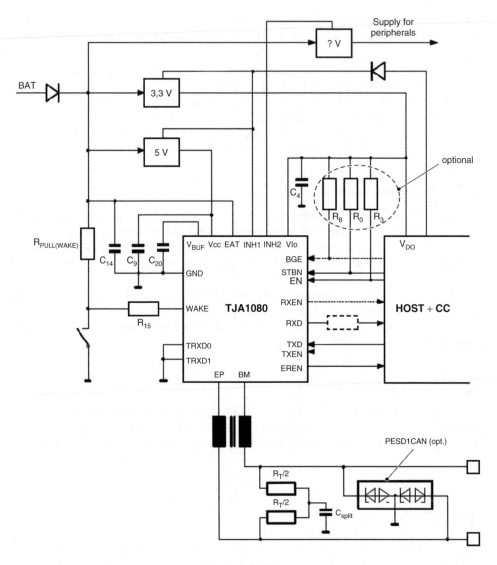

Figure 18.18

   To limit all these problems and their consequences, the specific Working Group of the FlexRay Consortium has published two very thick documents (each about 800 pages), one about the protocol (FlexRay Communications System Data Link Layer Conformance Test Specification) and the other about the physical layer (FlexRay Physical Layer Conformance Test Specification). They both describe, in great detail, all the procedures and procedural options for tests to be carried out before judging the conformity of the protocols and physical layer to the FlexRay specification 2.1 Rev. B and 3.0 – at 10 Mbit/s only, because that is what the specification says. It should be noted that development, validation, characterisation, and so on of satisfactory measurement testbeds for these tests are very long and very expensive. Warning to amateurs!

Independent, recognised laboratories such as C & S, TZM and TÜV NORD Mobilität GmbH & Co., KG Institute for Vehicle Technology and Mobility are approved, accredited and authorised by the Consortium to carry out these tests. As you certainly know, the axe of this type of procedure falls after a few hundred tests in the form of the famous rubber stamps 'Passed' or 'Failed', and as usual, if a single one of the test points does not conform, you simply do not conform and are rejected! It is a fact that in a network, different components from different suppliers must cohabit, and it is necessary to guarantee absolutely the functional interoperability of the network, to at least 110% ☺.

In short, only components which meet these tests are really worthy of carrying the FlexRay logo, and consequently are not subject to licences or royalties since they correspond to the definition of the rules stipulated by the Consortium. In the opposite case, they cannot have the FlexRay label, and if they use many FlexRay tricks (meaning patents, and so on), they then fall under the applications (and expense) of licences and royalties due to the Consortium. So be careful about 'almost' FlexRay components.

As an example, Figures 18.19 and 18.20 show facsimiles of the first two certificates of conformity, one issued for a Freescale microcontroller, the other for a Philips/NXP line driver.

18.5 Bus Guardian

One of the principal functions of the bus guardian is to prevent what is customarily called the 'babbling idiot' of the bus, and to prevent access to the medium by a node at a bad slot time. Its task consists of authorising or forbidding control by the line driver, detecting errors and supervising access to the medium. To do this, the bus guardian must be synchronised with the communication controller, know exactly the timing of communications and have an independent clock.

Several schools are contending to fulfil this function:

- in hardware
  - centralised bus guardians
  - distributed bus guardians
- in software.

For a long time, the Consortium has been keeping preliminary documents about the first two on the back burner, without ever finalising them as far as we know.

The functional block diagram of the combination of CC, line driver and bus guardian is shown in Figure 18.21.

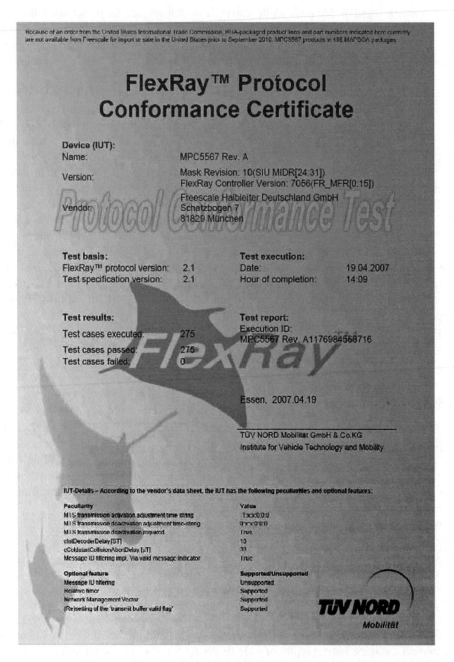

Figure 18.19

Authentication on
FlexRay Conformance

NXP Semiconductors Netherlands B.V.
Log. Inbound   BF-0.114
Gerstweg 2
6534 AE Nijmegen
Netherlands

T7 Mikroelektronik
Robert-Bosch-Str. 6
D-73037 Göppingen
Prof. Dr.-Ing. J. van der List

C&S 🖂
Fachhochschule
University of Applied Sciences
C&S – communication &
systems group
Prof. Dr. W. Lawrenz

Conformance Test Results

FlexRay component / part number	NXP, TJA1080A
Executed by	TZ Mikroelektronik – Robert-Bosch-Str. 6 – D-73037 Göppingen Tel.: +49 7161 5023-0 – Fax: +49 7161 5023-444 – info@tzm.de
Date of test	November 2007
Version of Test Specification	FlexRay Physical Layer Conformance Test Specification Version V2.1 Rev.A
Types of test	FlexRay Bus Driver with Active Star mode functionality and with communication controller interface

FlexRay Physical Layer
Conformance Test

Passed / Failed
Dynamic Test Cases: 100%   PASS *) **)
Static Test Cases: 100%   PASS

TZM and C&S are worldwide recognized as independent testing experts of FlexRay communication systems and FlexRay Transceivers.

Herewith TZM and C&S are proud to confirm that the test on the above specified device implementations have been performed.

Göppingen, 23rd of November 2007

*) 6.3.11.1-3: tested with wake-up application
**) 6.3.15.3, 6.3.7.1-2, 7.3.7.1-2: INH1 not checked, because no INH comparator on test hardware for V2.1 Rev.A

Edgar Grundstein
General Management, TZM

Daniel Holzy
Manager FlexRay &
Automotive Bussystems, T7M

Frank Fischer
Senior Engineer, C&S group

David Bollati
Projectmanager, C&S group

Headquarter: Steinbeis GmbH & Co. KG für Technologietransfer
Haus der Wirtschaft, Willi-Bleicher-Str. 19, D-70174 Stuttgart, Germany
Phone: +49 -7 11-18 39-5, Fax: +49-7 11-2 26 10 76, Internet: www.stw.de, E-Mail: stw@stw.de
Register court Stuttgart, HRA 12 480 General partner: Steinbeis Verwaltungs-GmbH, Register court: Stuttgart HRB 18715
Directors: Prof. Dr. Heinz Trasch (chairman), Prof. Dr. Michael Auer (deputy chairman), Prof. Dr.-Ing. Sylvia Rohr
A company of Steinbeis-Stiftung

Figure 18.20

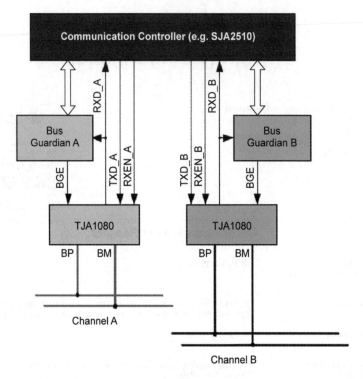

Figure 18.21

With these few words on the bus guardian, which can only evolve, that is the end of this quick presentation of the normal components that exist on the market on a given date. In the same way as for CAN, these product families will be daily enriched with new elements. We therefore leave it to you to take the trouble to visit regularly the catalogues of the various component manufacturers, to keep abreast of developments.

19

Tools for Development, Integration, Analysis and Testing[1]

As we have shown in numerous earlier chapters, networks operating under the FlexRay protocol are complex systems, and implementing them requires specific tools to help with design, testing and final integration.

19.1   The V-Shaped Development Cycle

Figure 19.1 shows the classic V-shaped cycle describing the various stages in the development of a system. To avoid speaking in a vacuum, and as an example, throughout this chapter we have indicated – superimposed on the cycle – different tools that are required during the various development and integration stages. These tools, the 'Da Vinci Tools' series, are marketed by one of the leaders in this sector, VECTOR.

This development series comprises three principal tools (see Figures 19.2 and 19.3):

- the 'System Architect';
- the 'Network Designer';
- the 'Developer'.

It constitutes a development environment which is dedicated to distributed systems, and supports all the modelling of the application, from the specification phase to the software integration phase, in the various ECUs of the system.

19.2   DaVinci Network Designer (Point 1 of the V Cycle)

Let us begin with the 'Network Designer'.

The first stage of implementing a FlexRay application consists of designing the FlexRay network; that is, implementing a communication system which includes several nodes which are directly connected to one or both of the two communication channels A and B.

---

[1] I must thank Mme Hassina Rebaïne, technical training manager at VECTOR France, for her help and great contribution to working out the content and editing this chapter.

FlexRay and its Applications: Real Time Multiplexed Network, First Edition. Dominique Paret.
© 2012 John Wiley & Sons, Ltd. Published 2012 by John Wiley & Sons, Ltd.

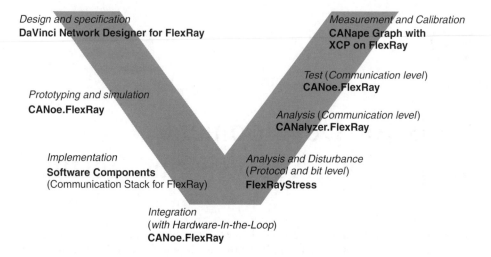

Figure 19.1 The V-shaped development cycle

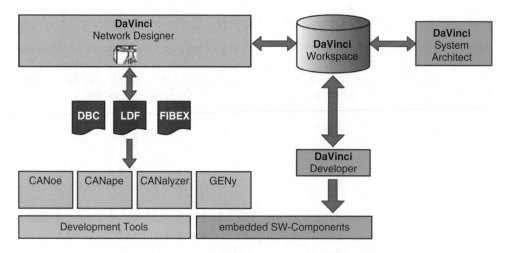

Figure 19.2 DaVinci Network Designer

This design makes it necessary to define the messages which one wants to transmit, their distribution and their scheduling on the communication bus, and thus to create a database. This phase is supported by the 'Network Designer' tool, which has a simple, user-friendly interface for configuring the nodes and defining their time windows, what is called a 'scheduler', the famous scheduler of the long appendices to Part B!

This tool is thus used to:

- **Model the network architecture and communication data** for distributed systems (who receives whom, who receives what and by whom, and so on, type of exchange).

The resulting description of data for FlexRay is generated in FIBEX (*.xml) format, which is defined by the Association for Standardisation of Automation and Measuring Systems (ASAM) standard.

- **Specify the parameters of the FlexRay protocol**, which are divided into two levels: the parameters at cluster level and the parameters at node level (see Figures 19.4 and 19.5).
- **Check the validity of the configuration** in relation to the FlexRay specifications, using the 'consistency check' function.

19.3   CANoe.FlexRay

Nowadays, the complexity of the communication systems in a vehicle makes a specific environment and methodology necessary to model, simulate, validate, test and do diagnostics on the virtual and real ECUs. Let us give some details about what is hidden behind these last two points.

19.3.1   Modelling, Simulation (Point 2 of the V Cycle)

What the manufacturers do to validate their systems is, firstly, to model and simulate the various logic controllers which will be put into the network (for example the values of the contents of the frames of the static segment, and so on) (see Figure 19.6).

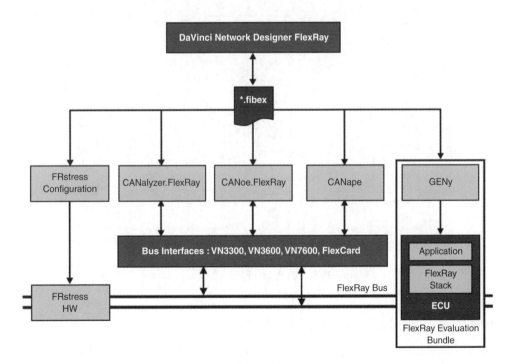

Figure 19.3

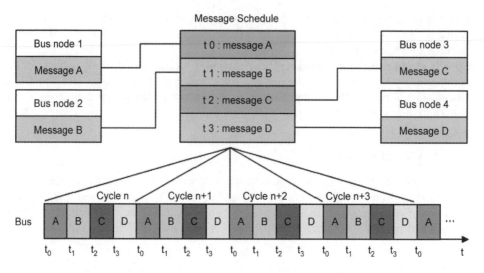

Figure 19.4 Defining and scheduling messages

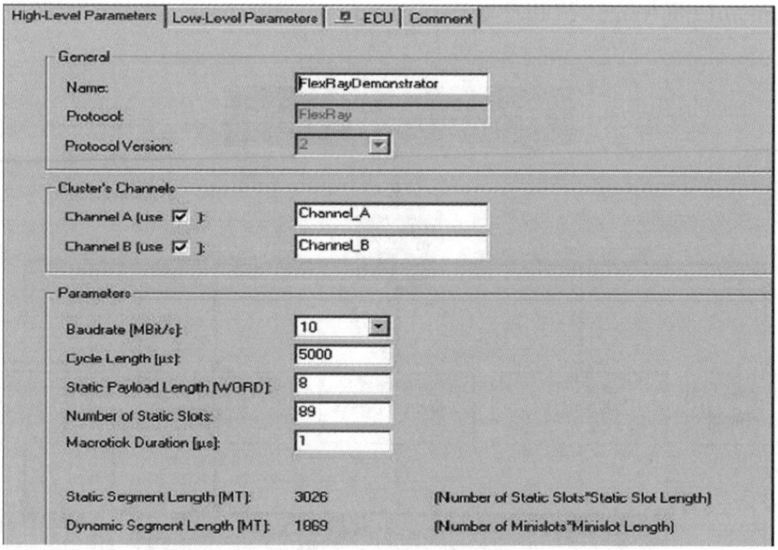

Figure 19.5 High-level FlexRay parameter

19.3.2   Integration (Point 4 of the V Cycle)

Next, stage by stage, the virtual logic controllers are replaced by the real components as they become available (phase 2 in Figure 19.7). Finally, the simulated environment is entirely replaced by the real logic controllers to be tested and validated in the final integration phase (see phase 3, also in Figure 19.7).

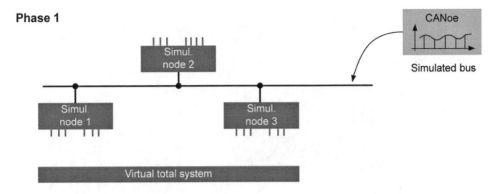

Figure 19.6   Hardware-in-the-loop: prototyping and simulation phase

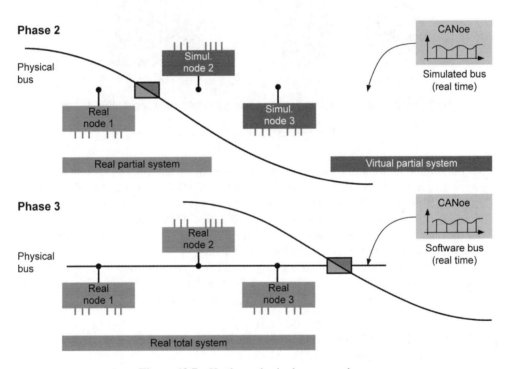

Figure 19.7   Hardware-in-the-loop: test phase

This approach is supported by the CANoe tool, which, by its hardware in the loop (HIL) methodology, offers the advantage of using a single simulation platform for the virtual (prototype phase), mixed (network using virtual and real nodes, integration phase) and finally real (test phase) network. In fact, the remaining (real and virtual) network and the entirely real network are simulated in the same environment as the entirely virtual network.

Additionally, CANoe offers the possibility of developing a simulation model in the form of 'panels', which make it possible to validate the overall system easily (see Figure 19.8).

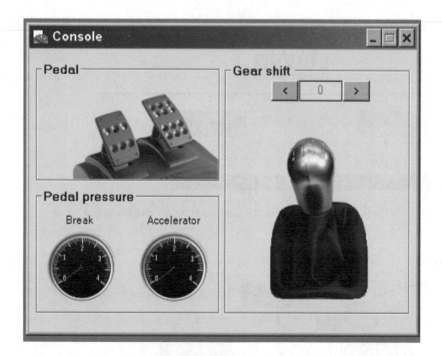

Figure 19.8   Simulation panels

In our example, the modelling of the simulated logic controllers/nodes is described using a proprietary event-driven language of the VECTOR company, called 'CAPL – Communication Access Programming Language', which is very close to the C language. Automatic generation tools, which are integrated with CANoe, make it possible, starting from the database, to generate these models and the corresponding panels, which enables the designer to validate the virtual architecture quickly.

In the phases described above (modelling, simulation and integration) and in order to display the frames, their content, their timing aspects, and so on it is useful or necessary to have a tool which makes it possible to display different windows, as CANalyser does.

19.4   FlexRay CANalyzer (Covers Points 2, 4 and 5 of the V Cycle)

The purpose of the special tool CANalyzer (an integral or separable part of CANoe) is to analyse the communications of the real logic controllers via various display windows: the graphic, data, statistics and trace windows.

For example, the trace window makes it possible to check exchanges of messages on the bus, their arrival times, the data they carry, and so on. Examples of windows are shown in Figures 19.9 and 19.10.

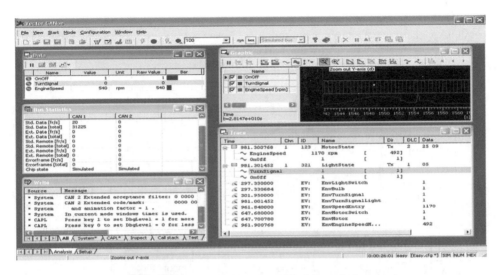

Figure 19.9   Analysis and measurement window

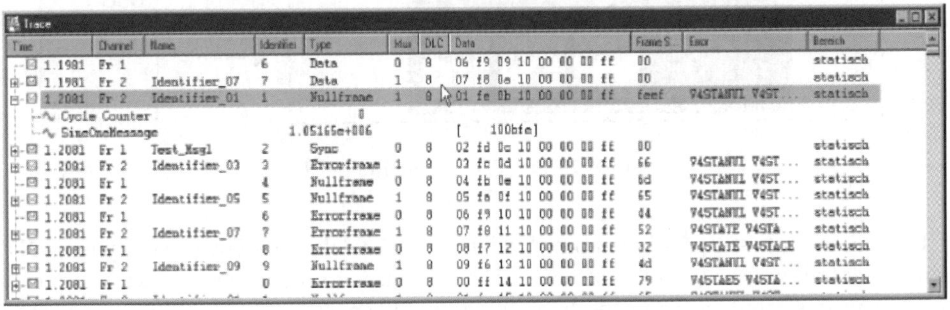

Figure 19.10   Trace window

19.5   Test and Diagnostics (Point 6 of the V Cycle)

The next phase, with all the real logic controllers on the network, corresponds to the test phase of the V cycle. To simplify the writing of testbeds with good coverage, CANoe is equipped with a test library (TSL, test set library), with functions making it possible to send 'stimuli' to the logic controller, and 'check' functions making it possible to check the validity of the expected results at any time. Thus, CANoe makes it easy to create test scripts via these functions and predefined test cases in the CAPL and XML languages. When these scripts have been executed by CANoe, a detailed test report is generated automatically in HTML format.

COMMENT

All the functions mentioned up to this point are summarised in Figure 19.11. As well as the modelling, simulation and test functions, CANoe also includes analysis and diagnostic functions in the same integrated environment, with simple, user-friendly interfaces (Figure 19.11).

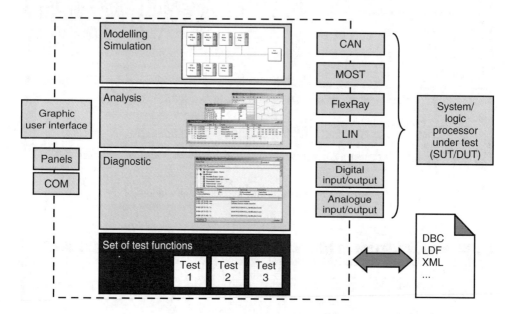

Figure 19.11   CANoe environment

Diagnostic requests, predefined in a database of CANdela Data Diagnostic type (CDD) can be sent via a user-friendly interface. The responses to these requests are interpreted directly in this interface or in the trace window. Diagnostic scripts can also be written in CAPL and XML (via functions of the test and diagnostic library) and executed by CANoe, with generation of a detailed report in HTML format. The KWP2000 and UDS protocols are supported. The memory of the logic controller can be read to recover the fault codes via the user interface.

19.6   Features of the FlexRay Protocol

The 'time-triggered' part of FlexRay communication requires a simulation platform which is capable of receiving and generating messages and signals in perfect synchronism with the scheduler which is provided in the communication network. CANoe manages this synchronism. Figure 19.12 shows two specific cases.

With the CANoe tool, the user has the possibility, via the modelling language CAPL, of acting on received and sent messages, on elapsed times (timers) or on errors. This concept has been extended to FlexRay, to act in synchronism at very specific 'time points' in the FlexRay cycle. This makes it possible to act, for example at the start of each cycle, at the end of a slot, and so on (see Figure 19.13). This synchronism makes it possible to monitor

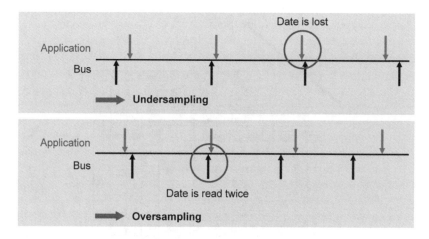

Figure 19.12 Example of undersampling and oversampling

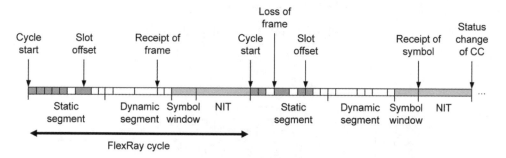

Figure 19.13 Example of notification and activation of 'time points' following FlexRay scheduler

times at the level of the communication medium, and to synchronise the application to the communication network.

On the other hand, the FlexRay signals, messages and frames are sent cyclically at high, different repetition rates. The 'cycle multiplexing' function is also supported by the CANoe simulator (see Figure 19.14).

In particular, thanks to the very low time of its operating system, its 'communication stack' and its 'runtime environment', which make it possible to guarantee very fast reaction times, CANoe is capable of supporting the mode of operation of 'response within the cycle' type.

CANoe also supports the FIBEX database format for analysing communications via analysis and measurement windows. The user can use the symbolic names of the signals and messages of the FIBEX database. The configuration parameters of the communication controller can also be read from FIBEX. It is also possible to send messages based on the scheduling table defined by, and in, DaVinci Network Designer.

As we indicated above, modelling simulated FlexRay nodes is described using the proprietary event-driven language CAPL. To simplify the writing of these models, CAPL

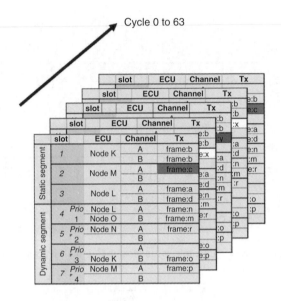

Figure 19.14   Cycle multiplexing

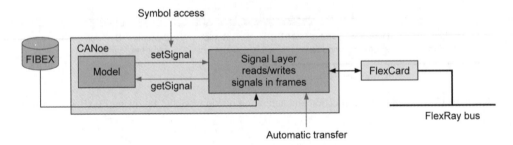

Figure 19.15   Signal layer: reading/writing frame signals

permits direct access to the FIBEX database, to recover the symbolic names of signals and messages. On the other hand, within FlexRay a buffer mechanism is implemented (signal layer), to activate reading and writing signals at a given instant (Figure 19.15). The frame which carries the thus-updated signal is sent and received automatically, at the predefined time point, in the scheduler of the FIBEX database.

19.7   Communication Interface

The models can also come from a user library (program written in another language) in dynamic link library (DLL) form. Via CANoe-API, these DLLs make it possible to access functions for reading or sending messages, signals or other things from the database. These DLLs are declared at the level of the CANoe user interface.

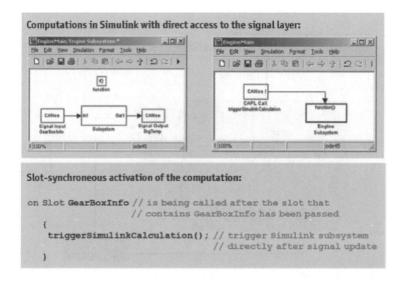

Figure 19.16 Matlab/Simulink in CANoe

Specific DLLs also make it possible to integrate 'MATLAB/SIMULINK' models into the CANoe environment. This enables the user to simulate complex models (Figure 19.16). The compiled code of CAPL models is executed directly at machine level.

IMPORTANT NOTE

If the machine (PC, etc.) is not really capable of execution in real time, this can result in interruptions of the simulation and to considerable, non-deterministic jitter on the communication bus.

19.7.1 CANoe Real Time

To meet strict real time constraints, 'CANoe Real Time' implements an architecture which uses two 'Dual PCs' connected by an Ethernet link (see Figure 19.17). This architecture divides the CANoe application into two parts: 'Soft Real-time Client', indicated on the diagram by 'CANoe GUI Client', for the user interface, and 'Real-time Server' for executing the models and their interface with the communication bus. The 'Real-time Server' is executed on a dedicated real time machine. CANoe Real Time supports various systems, Windows XP, Windows XP Embedded and Windows CE, which can offer task reaction times of less than 10 μs (depending on which bus interface is used).

This architecture provides the following advantages:

- guarantee of 100% processing of the FlexRay network load;
- very short response times;
- simultaneous simulations of several ECUs;
- precise treatment of synchronism between the communication bus and the application;
- responses within the cycle;
- constant, minimal jitter.

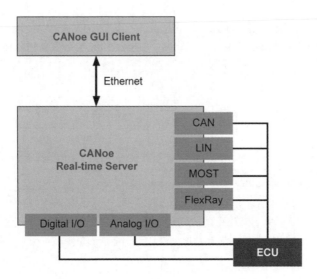

Figure 19.17   Dual PC architecture of CANoe

For the interface bus, various formats are possible: PCMCIA with the FlexCard, PCI with the VN3300, USB with the VN3600 and, today, VN7600 for FlexRay in USB connections.

19.7.2   FlexRayStress

FlexRayStress makes it possible to simulate errors on the communication network and to test the consequent behaviour of the ECUs. It makes it possible not only to disturb the communication medium by, for example, simulating a short circuit, but also to disturb transmission of messages from nodes on the network.

19.7.3   CANape FlexRay

Once the logic controller is implemented in the vehicle, the task of the application engineer consists of integrating this logic controller into the system and adapting it to the environment. This task requires optimisation of several parameters, for example optimal control of the algorithm of a function in the system. These parameters can be accessed by hardware or software, via a standard ASAM interface.

We have now summarised in a few paragraphs the content of one example of a set of necessary tools for development, implementation, diagnostics and tests of an application which supports the FlexRay protocol.

20

Implementation of FlexRay Communication in Automotive Logic Controllers[1]

20.1 FlexRay and AUTOSAR

As we have indicated before, networks functioning under the FlexRay protocol are complex systems, and when a distributed system of logic controllers interconnected by a FlexRay network is implemented, the development of the software and the implementation of these logic controllers become very important. Apart from the operational functions (the application software), the software must ensure, in particular, that information is exchanged between the logic controllers while complying with the communication rules of the network. While the communication controller is responsible for a large part of this task, it must nevertheless be configured and implemented by the software, to fulfil the communication requirements of the application software. This is the job of what is now called the communication stack, because this software is structured in several stacked modules.

When the moment came to implement the first FlexRay communication stacks, a revolution was under way in the automotive software environment. The vehicle manufacturers were persuaded by the increased complexity of electronic systems and software to initiate a partnership to improve their mastery of this complexity. Automotive Open System Architecture (AUTOSAR) has thus defined a development methodology and a standard architecture for the software of automotive logic controllers. The communication stack has thus been standardised, covering the FlexRay network in particular. The implementation of FlexRay communication in the software of automotive logic controllers has thus become completely linked to the AUTOSAR work.

After a presentation of the AUTOSAR partnership, in this chapter we will describe communication between logic controllers in an AUTOSAR system, and more particularly

---

[1] I must thank Mr Jean-Philippe Dehaene, technical director at VECTOR France, for his great contribution to working out the content and editing this chapter.

FlexRay and its Applications: Real Time Multiplexed Network, First Edition. Dominique Paret.
© 2012 John Wiley & Sons, Ltd. Published 2012 by John Wiley & Sons, Ltd.

AUTOSAR communication on a FlexRay network. We will then have described the
AUTOSAR FlexRay communication stack.

However, AUTOSAR does not currently cover all the communication requirements of
automotive logic controllers. Reprogramming the logic controllers, adjusting the algorithms, uses specific communications which are not fully implemented by the AUTOSAR
specification. We will therefore also describe these additional needs and how they are
implemented in the software.

20.2   The AUTOSAR Partnership

The AUTOSAR partnership was initiated in 2002 by BMW, Daimler, VW with Bosch,
Continental and Siemens, and subsequently joined by Ford, PSA, Toyota and finally GM.
Its purpose is greater mastery of electronic and software developments for the motor
vehicle. This includes, in particular:

- standardising the basic functions that all manufacturers require;
- decoupling the implementation of application software from what logic controller is
  used;
- enabling it to be transferred easily from one logic controller to another;
- thus permitting economies of scale according to the different variants of vehicle platform;
- permitting easy integration of software from different suppliers in one logic controller;
- meeting the increased requirements for availability and security;
- guaranteeing 'maintainability' and allowing updates and additions throughout the life
  cycle of the vehicles;
- and so on.

After a phase of definition of principles, the 'core partners' surrounded themselves with
'premium members' from other automotive manufacturers, equipment manufacturers of
the first rank, developers of embedded software and software tools and partners from the
microcontroller industry.

From September 2004, numerous work groups have been launched, making it possible
to create the first versions of the detailed specifications of the standard. Implementation of
prototypes then made it possible to test these specifications, which were then improved.
Version 3.0, published at the end of 2007, reached a sufficient level of maturity to allow
the launch of mass production vehicles which implemented the AUTOSAR standard on
a large scale. However, the work of the partnership continued, and a greatly enriched
version 4.0 was published in December 2009, and revised in April 2011.

20.3   Communication in an AUTOSAR System

AUTOSAR makes it possible to control the definition and implementation of a distributed
system. This standard thus plays a large part in the support of communications that
appear in such a system. Following the AUTOSAR methodology, we will show how
these communications are implemented.

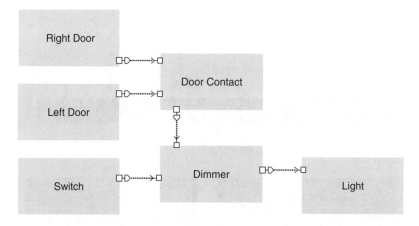

Figure 20.1

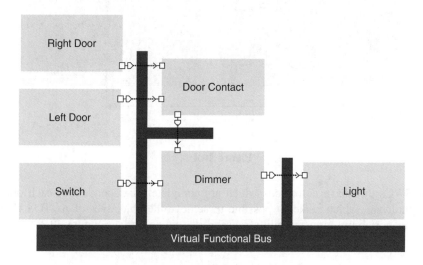

Figure 20.2

20.3.1 Functional Analysis, Virtual Function Bus

The AUTOSAR methodology is based on functional analysis of the functions to be implemented. The functions are divided into so-called software components (SWCs), which communicate with each other to make the service global (see Figure 20.1).

These SWCs are defined independently of any hardware architecture. It is possible to represent exchanges of data in an equivalent way to exchanges of information on a communication bus; this is one of the fundamental concepts of AUTOSAR, the virtual functional bus (VFB) (see Figure 20.2).

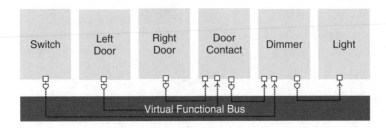

Figure 20.3

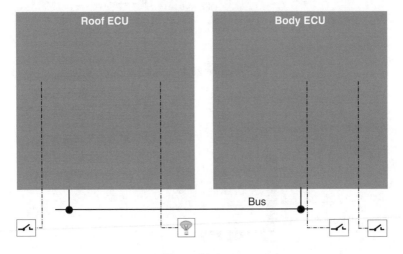

Figure 20.4

The SWCs then have definitions which are completely independent of each other and of the hardware, which makes it possible to achieve the objectives of AUTOSAR in the matters of portability, transferability, reusability, and so on (see Figure 20.3).

20.3.2 Passing from Virtual to Real

While defining the concepts of SWC and VFB makes it possible to obtain the qualities that the AUTOSAR partnership wanted, a real system cannot be content with virtual concepts. It will consist of logic controllers (electronic control units, ECUs) which are interconnected by networks. It is therefore right to describe this hardware architecture and to decide how it takes responsibility for the different SWCs (see Figures 20.4 and 20.5).

Virtual communication via the VFB must now be really implemented in each ECU, and on the networks that interconnect them. To do this, a procedure in several stages is adopted.

Firstly, the VFB is implemented in each ECU by a fundamental component of the AUTOSAR standard: the run time environment (RTE).

Within a logic controller, the RTE provides exchange of data between the SWCs which are integrated in this logic controller, exchange of data between these SWCs and the

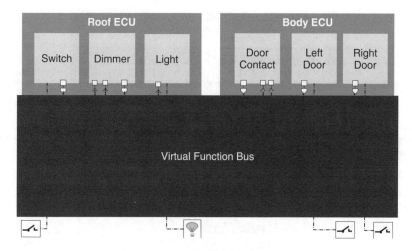

Figure 20.5

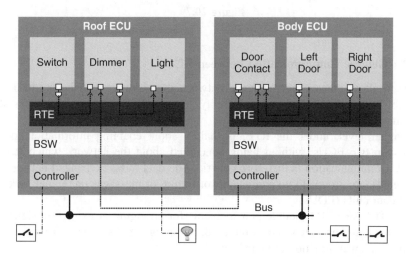

Figure 20.6

sensors or actuators and exchange of data with the remote SWCs via a communication stack (see Figure 20.6).

The RTE is based on basic software (BSW), which implements access to the hardware resources of the ECU in 'drivers'. The communication stacks on the networks are part of the BSW.

20.3.3   AUTOSAR FlexRay Communication Stack

Figure 20.7 gives an overall view of the AUTOSAR communication stack for a FlexRay network. We will give details of the different parts in the following pages.

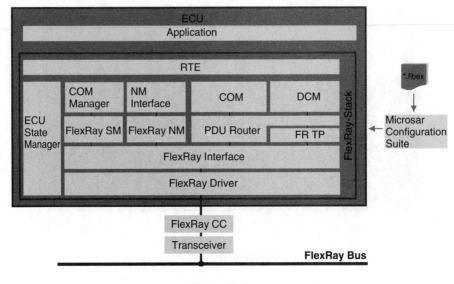

Figure 20.7

20.3.3.1 Generic Communication between ECUs

In order to allow greater independence between the software and the hardware, which guarantees better portability and reusability, the AUTOSAR communication stack is divided into two levels.

The higher level, under the RTE, is mainly about enabling information to be carried between two ECUs, without being concerned about the network which is used for transport. Communication is therefore generic. The data to be exchanged between ECUs, called signals, are grouped into equivalents of frames or messages on the networks, called protocol data units (PDUs).

The AUTOSAR COM module is responsible for arranging or extracting signals in the PDUs, sequencing the PDUs and, if required, rerouting signals directly from one network to another (gateway function at signal level).

The AUTOSAR COM module depends on the AUTOSAR PDU Router module, which routes the PDUs on the network concerned and, if required, reroutes a PDU directly from one network to another (gateway function at PDU level).

An additional module, the IPDU multiplexer, makes it possible, if necessary, to multiplex the content of a PDU, which can then carry several different configurations of signals.

20.3.3.2 Communication on a FlexRay Network

AUTOSAR has standardised communication on three types of automotive network: CAN, LIN and FlexRay.

Again, to guarantee the greatest possible independence between the software and the hardware, this communication is divided into two layers.

The higher layer, called the interface, implements communication on the network independently of the technology of the microcontroller and communication controller, whether it is internal or external to the microcontroller.

In the case of FlexRay, this is the AUTOSAR FlexRay interface module, which ensures that PDUs are assembled into FlexRay frames, manages the update bits of the PDUs and notifies the higher layers that PDUs have arrived or been successfully sent.

It depends on the lower, hardware-dependent layer, called the micro controller abstraction layer, either to communicate with an external communication controller or to manage the internal communication controller.

In this second case, when it is called up so that FlexRay can be generalised with the ever-greater integration of the hardware, it involves the AUTOSAR FlexRay driver module, which is responsible for initialisation of the communication controller, transmission and reception of FlexRay frames and detection of errors returned by the communication controller.

20.3.3.3 Coordination of the Network

AUTOSAR also standardises how the ECUs are coordinated on the networks, in particular to take account of how the system is started and stopped.

A first AUTOSAR module, the ECU state manager (SM), initialises all the BSW of the ECU and the logic of changing operating modes of the system (Stop, Go, Standby) and transient modes (Start, Switchoff).

It depends on a second module, which is responsible for managing the states of the networks, the communication manager. It initialises the communication modules, starts and stops communication on the networks and manages communication faults.

For the FlexRay network, this module depends on the FlexRay SM module, which is responsible for waking up and starting this network.

The various ECUs on the network communicate with each other in a network management (NM) function, to coordinate with each other and decide on stopping the network. This function is provided generically by the AUTOSAR NM interface module, which for coordinating the FlexRay network specifically depends on a FlexRay NM module.

20.3.3.4 Communication with Diagnostic Tools

For communication with the diagnostic tools, AUTOSAR is based on the ISO 14229 Unified Diagnostic Services standard. This protocol is implemented in the AUTOSAR diagnostic communication manager module, which is directly connected to the PDU router module.

The messages which are exchanged by this protocol can be longer than the size of a PDU. In fact, for reasons of compatibility with CAN and LIN networks, the size of PDUs can typically be fixed at 8 bytes, whereas the diagnostic protocol sometimes involves exchanges on a large scale, in particular for reading the fault memory or for reprogramming logic controllers.

Communication must then be segmented. For FlexRay, this function is based on the ISO 15765-2 and 15765-4 standards, and is the responsibility of the AUTOSAR FlexRay transport protocol (TP) module.

Appendix of Part E

This appendix, which is presented in the form of tables (Tables E.1–E.3), is not intended to be comparative or a list of limitations, properties, potential qualities, and so on, but has as its sole purpose to indicate and summarise the principal structural and inherent differences between the two protocols CAN (and its variant TTCAN) and FlexRay, which were each developed for distinct purposes. There is therefore no conflict between CAN and FlexRay, but, on the contrary, complementarity.

Table E.1

Parameters		CAN	TTCAN	FlexRay
Protocol	–	–	–	–
Specification reference	–	ISO 11 898 1/2/3	ISO 11 898-4	FlexRay 3.0
Communication	–	Asynchronous	Asynchronous	Asynchronous
	–	Broadcast/ multicast	Broadcast/ multicast	Broadcast/ multicast
Bit rate (gross)	–	Constant	Constant	Constant
	Mbit/s	LS➔0–0.125 HS➔0.125–1	LS➔0–0.125 HS➔0.125–1	10–5–2.5 –
Bit	Duration	HS minimum = 1 μs	HS minimum = 1 μs	Nom = 100 ns
	Coding	NRZ	NRZ	NRZ
	Logical '0'	Dominant	Dominant	Dominant
	Logical '1'	Recessive	Recessive	Dominant
	'Idle'	Recessive	Recessive	Recessive
	–	Bit stuffing	Bit stuffing	n/a
Production of a communication on the network	Of type	Spontaneous (event-triggered)	Cyclical by operating cycles	Cyclical by operating cycles
	Static segment	–	–	Time-triggered
	Dynamic segment	–	–	Event-triggered

(continued overleaf)

FlexRay and its Applications: Real Time Multiplexed Network, First Edition. Dominique Paret.
© 2012 John Wiley & Sons, Ltd. Published 2012 by John Wiley & Sons, Ltd.

Table E.1 (*continued*)

Parameters		CAN	TTCAN	FlexRay
Real time aspect	–	No/badly	Possible/almost	Yes
	Deterministic	No	–	Yes – in the static segment
	Probabilistic	Yes	Probable in deter	Yes – in the dynamic segment
Access to medium	–	Via bit arbitration during UID	Via slot number and bit arbitration during arbitration	Via hierarchy of priorities designed offline
Arbitration	–	CSMA	TDMA/CSMA	TDMA
	Static segment	–	–	TDMA time slot
	Static segment	–	–	FTDMA mini time slots
	Time multiplexing	–	–	Via cycle numbers
Fault tolerant	–	HS no	–	… Small yes 'never give up'
	–	LS yes	–	–
Data format	–	Byte = 8 bits	Byte = 8 bits	Byte = 10 bits coding 8N1
Frame length	–	0–8 bytes	0–8 bytes	127 words of 16 bits = 254 bytes
CRC	–	–	–	After header 11 bits
				At end of frame 24 bits
Useful bandwidth Net bit rate	Mbit/s	Maximum approximately 0.5–1 Mbit/s gross	Maximum approximately 0.5–1 Mbit/s gross	Maximum approximately 7 M per channel @ 10 Mbits/s
				Double if complementary data on two channels
Redundancy	–	Badly	–	Yes, possible
	Of data	–	–	Two distinct transmission channels
	Physical	–	–	Two distinct transmission channels

(*continued overleaf*)

Table E.1   (*continued*)

Parameters		CAN	TTCAN	FlexRay
Physical layer	–	–	–	–
Medium	Wire	Yes	Yes	Yes
	Optical fibre	Yes	Yes	Yes
Communication channels	–	Single	Single	Single/double
Topologies	Linear bus	Yes	Yes	Yes
	Linear bus + stub	Yes	Yes	Yes
	Passive star	Yes	Yes	Yes
	Active star	Tricky	Tricky	Yes
	Repeater	Tricky	Tricky	Yes
	Hybrid	Tricky	Tricky	Yes
	Ring	Tricky	Tricky	Yes
Taking account of elements present on the medium	Length/speed of propagation	Nothing (see Section synchro)	–	Compensation for propagation delays
	Turnaround speed	Nothing	–	Truncation management (TSS)
Synchronisation between participants	–	Device for re-synchronising sampling point	–	Via creation of a 'Global Time' of the network
	In bit rate	Via 'phase segments'	–	Adjustment of rate during the NIT
	In phase	Via 'phase segments'	–	Adjustment of offset during the NIT

NRZ = no return to zero; TDMA = time division multiple access; FTDMA = flexible time division multiple access; NIT = network idle time.

Table E.2

Criteria	CAN	FlexRay
Bandwith	Busload on different CAN sub-buses at limit	Net data rate 5 MBit/s at gross 10 MBit/s
	Multiple gateways cause unacceptable delays	Flexible use of bandwidth
Topologies	CAN requires bus with dominant/recessive state	Star topology provides improved electrical characteristics and fault isolation
		Flexible topologies

(*continued overleaf*)

Table E.2   *(continued)*

Criteria	CAN	FlexRay
Deterministic communication	Non-deterministic behaviour of CAN at high bus loads results in poor quality of service	Deterministic latencies (guaranteed transmission time for frames in static segment)
System integration	System integration can cause strange side effects by increasing bus load	System integration does not change any timing
Fault tolerance	Underlying operating concept does not consider application level fault tolerance	Underlying operating concept considers application level fault tolerance (redundant channel, fault-tolerance clock synchronisation)
Application level replica determinism	Synchronisation of application tasks requires additional communication	Synchronisation of application tasks through synchronised time base

Table E.3

Feature	CAN	TTP	Byteflight	FlexRay
Message transmission	Asynchronous	Synchronous	Synchronous and asynchronous	Synchronous and asynchronous
Message identification	Message identifier	Time slot	Message identifier	Time slot
Date rate	1 MBit/s gross	2 MBit/s gross	10 MBit/s gross	10 MBit/s gross
Bit encoding	NRZ with bit stuffing	Modified frequency modulation (MFM)	NRZ with START/ STOP bits	NRZ with STAR/ STOP bits
Physical layer	Transceiver up to 1 MBit/s	Not defined	Optical transceiver up to 10 MBit/s	10 MBit/s with differential signalling
Clock synchronisation	Not provided	Distributed, in microsecond range	By master, in 100 ns range	Distributed, in microsecond range
Temporal composability	Not supported	Supported	Supported for high priority messages	Supported
Latency jitter	Bus load dependent	Constant for all messages	Constant for high priority messages according to t_cyc	Constant for all messages

(continued overleaf)

Table E.3   (*continued*)

Feature	CAN	TTP	Byteflight	FlexRay
Error containment	Not provided	Provided with special physical layer	Provided by optical fibre and transceiver	Provided with special physical layer
Babbling idiot avoidance	Not provided	Only by independent bus guardian	Provided via star coupler	Provided via star coupler or bus
Extensibility	Excellent in non-time critical applications	Only if extension planned in original design	Extension possible for high priority messages with effect on bandwidth	Separation of functional and structural domain
Flexibility	Flexible bandwidth for each node	Only one message per node and TDMA cycle	Flexible bandwidth for each node	Multiple slots per node, dynamic

21

Conclusion

We have now arrived at the end of this work, which is dedicated to the technical presentation of the broad outline of the FlexRay protocol.

As you will have noticed, compared with CAN, there is great similarity and great parallelism of technical and strategic approaches in how this new concept was worked out and developed. Also, it is practically the same players who initiated the process of designing and industrialising FlexRay and CAN, with the same proactive emphasis and great determination. Similarities are nice sometimes . . . ! All that is left to do is to fill in, day by day, for the next two decades, a table showing the introduction of vehicles onto the market and it will all be over!

Perhaps, in the course of this period, because of having to solve problems associated with ever more and ever bigger redundant safety equipment, we shall also see a technical 'merger' between the two markets, for the motor vehicle and aeronautics, which do not compete commercially. That would be magnificent; we would be going back to the common origins of the motor vehicle and civil aviation (Voisin, Hispano, and so on).

While we're waiting, we hope at least that this book will have given you new ideas for your future systems, and we now make an appointment with you for about 2015 to 2017 for the first comments, taking care to be at a certain distance from the introductory phase of this system.

One last point: as you will certainly have noticed since about mid-2007, the FlexRay protocol is practically stable, and all its limitations have been known since the date of its official publication. In principle, therefore, the people who presided over its development within the Consortium should be laid off . . . or almost! After this joke, obviously they aren't! As usual, we have consulted our favourite crystal ball, which has again given us some indications:

- Firstly, although it may seem surprising to some people, it is true that the Consortium has, in principle, finished its work and has dissolved itself.
- There is then the problem of maintaining the protocol through the years. This problem is usually solved by 'passing the buck' to the appropriate standardisation authorities of ISO, which themselves have known maintenance (amendment) cycles and periodic revisions – at least every five years – of the standards (as was the case with the R. Bosch company for CAN, with the ISO 11 898 – x series of standards). That should

---

FlexRay and its Applications: Real Time Multiplexed Network, First Edition. Dominique Paret.
© 2012 John Wiley & Sons, Ltd. Published 2012 by John Wiley & Sons, Ltd.

also make it possible, on the same occasion, to resolve officially how to open up FlexRay applications to 'non-automotive' applications.

• Finally, don't worry about the young unemployed people mentioned a few lines earlier. They have already begun work on the next multiplexed networks for the coming decades! Phew, we were worried! After FlexRay is opened up to 'X-by-Wire', will we go to 'X-by-Wireless'? Probably, but we will have more than 20 years to wait. Will we be able to endure this unbearable suspense for so long?

After these terrifically encouraging statements, we hope to meet you again soon – or quite soon – on these subjects!

Appendix 1

The Official Documents

Since this book does not claim to summarise everything or to be exhaustive on the subject, to satisfy your curiosity and your knowledge, you will find below, as of 1 January 2010, the list of documents to be found on the website of the FlexRay Consortium, www.flexray.com/. We warmly advise and recommend that you download them (... and open and read them, naturellement) to discover the numerous details that we have, more or less willingly, failed to present to you! Bear in mind, however, that, as outlined in Chapter 16, v3.0 of the specifications have now been published by the Consortium. These do not appear on the website for download and must be purchased.

- General
  - Requirements Specification v2.1
- Protocol
  - Protocol Specification v2.1 Rev. A
  - Protocol Specification v2.1. Rev. A Errata v1
- Physical layer
  - Electrical Physical Layer Specification v2.1 Rev. B
  - Electrical Physical Layer Application Notes v2.1 Rev. B
  - Electrical Physical Layer Spec. v2.1 Rev. B Errata sheet v2.0
  - Physical Layer EMC Measurement Specification v2.1
  - Physical Layer Common Mode Choke EMC Evaluation Specification v2.1
- Bus guardian
  - Preliminary Node-Local Bus Guardian Specification v2.0.9
  - Preliminary Central Bus Guardian Specification v2.0.9
- Conformity tests – CT
  - FlexRay Protocol CT Specification v2.1.2
  - FlexRay Data Link Layer CT Specification v2.1.1
  - Electrical Physical Layer CT Specification v2.1 Rev. A
  - Electrical Physical Layer CT Specification v2.1 Rev. B
  - FlexRay Physical Layer CT Specification 2.1 Rev. A Errata Sheet 2
  - FlexRay Phy_CT Specification 2.1 Rev. B Heterogeneous Tests public.pdf

FlexRay and its Applications: Real Time Multiplexed Network, First Edition. Dominique Paret.
© 2012 John Wiley & Sons, Ltd. Published 2012 by John Wiley & Sons, Ltd.

Appendix 2

Principal Parameters of the FlexRay Protocol

The parameters presented in this appendix are from the FlexRay specification release v 2.1B, and in a few pages make it possible to summarise the points to be observed when a project is being developed.

They are divided into three large branches:

- the protocol constants, which fix the principal entities of your application;
- the aspect dedicated to the global parameters of the application;
- the aspect of the specific parameters of the node.

Table A2.1 Constants of the FlexRay protocol

Name	Description	Range	FIBEX
CCASActionPointOffset	Initialisation value of the collision avoidance symbol (CAS) action point offset timer.	1 MT	n.a.
cChannelIdleDelimiter	Duration of the channel idle delimiter.	11 gdBit	n.a.
cClockDeviationMax	Maximum clock frequency deviation, equivalent to 1500 ppm (1500 ppm = 0.0015).	0.0015	n.a.
cCrcInit[A]	Initialisation vector for the calculation of the frame CRC on channel A (hexadecimal).	0xFEDCBA	n.a.
cCrcInit[B]	Initialisation vector for the calculation of the frame CRC on channel B (hexadecimal).	0xABCDEF	n.a.
cCrcPolynomial	Frame CRC polynomial (hexadecimal).	0x5D6DCB	n.a.
CcrcSize	Size of the frame CRC calculation register.	24 bits	n.a.

(continued overleaf)

FlexRay and its Applications: Real Time Multiplexed Network, First Edition. Dominique Paret.
© 2012 John Wiley & Sons, Ltd. Published 2012 by John Wiley & Sons, Ltd.

Table A2.1   *(continued)*

Name	Description	Range	FIBEX
cCycleCountMax	Maximum cycle counter value in a cluster.	63 cycles	n.a.
CdBSS	Duration of the byte start sequence.	2 gdBit	n.a.
CdCAS	Duration of the logical low portion of the collision avoidance symbol (CAS) and media access test symbol (MTS).	30 gdBit	n.a.
cdCASRxLowMin	Lower limit of the CAS acceptance window.	29 gdBit	n.a.
CdCycleMax	Maximum cycle length.	16 000 µs	n.a.
CdFES	Duration of the frame end sequence.	2 gdBit	n.a.
CdFSS	Duration of the frame start sequence.	1 gdBit	n.a.
cdMaxMTNom	Maximum duration of a nominal macrotick. Each implementation must be able to support macroticks of up to this length. Different implementations may support higher values.	6 µs	n.a.
cdMinMTNom	Minimum duration of a nominal macrotick. Each implementation must be able to support macroticks of at least this length. Different implementations may support lower values.	1 µs	n.a.
cdTxMax	Longest possible period of continuous transmission activity for a valid FlexRay configuration.	1433 µs	n.a.
cdWakeupMaxCollision	Number of continuous bit times at LOW during the idle phase of a WUS that will cause a sending node to detect a wakeup collision.	5 gdBit	n.a.
cdWakeupSymbolTxIdle	Duration of the idle phase between two low phases inside a wakeup pattern.	18 µs	n.a.
cdWakeupSymbolTxLow	Duration of low phase of a transmitted wakeup symbol.	6 µs	n.a.
cHCrcInit	Initialisation vector for the calculation of the header CRC on channel A or channel B (hexadecimal).	0x01A	n.a.
cHCrcPolynomial	Header CRC polynomial (hexadecimal).	0x385	n.a.
cHCrcSize	Size of header CRC calculation register.	11 bits	n.a.
cMicroPerMacroMin	Minimum number of microticks per macrotick during the offset correction phase.	20 µT	n.a.
cMicroPerMacroNomMax	Maximum number of microticks in a nominal (uncorrected) macrotick.	240 µT	n.a.
cMicroPerMacroNomMin	Minimum number of microticks in a nominal (uncorrected) macrotick.	40 µT	n.a.

Table A2.1 (*continued*)

Name	Description	Range	FIBEX
cPayloadLengthMax	Maximum length of the payload segment of a frame.	127 words	n.a.
cPropagationDelayMax	Maximum propagation delay from the falling edge (in the BSS) in the transmit signal of node M to corresponding falling edge at the receiver of node N.	2.5 µs	n.a.
cSamplesPerBit	Number of samples taken in the determination of a bit value.	Eight samples	n.a.
cSlotIDMax	Highest slot ID number.	2047 slots	n.a.
cStaticSlotIDMax	Highest static slot ID number.	1023 slots	n.a.
cStrobeOffset	Sample where bit strobing is performed (first sample of a bit is considered as sample 1).	Five samples	n.a.
cSyncNodeMax	Maximum number of sync nodes in a cluster.	15 nodes	n.a.
cVotingDelay	Number of samples of delay between the RxD input and the majority voted output in the glitch-free case.	Two samples	n.a.
cVotingSamples	Numbers of samples in the voting window used for majority voting of the RxD input.	Five samples	n.a.

WUS = wakeup symbol; BSS = byte start sequence.

Table A2.2 Global parameters

Name	Description	Range	FIBEX
gAssumedPrecision	Assumed precision of the application network.	0.15–11.7 µs	n.a.
gChannels	The channels that are used by the cluster.	(A, B, A and B)	CHANNEL-REFS
gClusterDriftDamping	The cluster drift damping factor, based on the longest microtick gdMaxMicrotick used in the cluster. Used to compute the local cluster drift damping factor pClusterDriftDamping.	0–5 µT	CLUSTER-DRIFT-DAMPING
gColdStartAttempts	Maximum number of times a node in the cluster is permitted to attempt to start the cluster by initiating schedule synchronisation.	2–31 times	COLD-START-ATTEMPTS

(*continued overleaf*)

Table A2.2   (*continued*)

Name	Description	Range	FIBEX
gdActionPointOffset	Number of macroticks the action point is offset from the beginning of a static slot or symbol window.	1–63 MT	ACTION-POINT-OFFSET
gdBit	Nominal bit time	8 × gdSample-ClockPeriod	BIT
gdBitMax	Maximum bit time taking into account the allowable clock deviation of each node.	gdBit × (1 + 0.0015) (μs)	n.a.
gdBitMin	Minimum bit time taking into account the allowable clock deviation of each node.	gdBit × (1−0.0015) (μs)	n.a.
gdCASRxLowMax	Upper limit of the CAS acceptance window.	67–99 gdBit	CAS-RX-LOW-MAX
gdCycle	Length of the cycle	10–16 000 μs	CYCLE
gdDynamicSlotIdle-Phase	Duration of the idle phase within a dynamic slot.	0–2 Minislots	DYNAMIC-SLOT-IDLE-PHASE
gdMacrotick	Duration of the cluster-wide nominal macrotick	1–6 μs	MACROTICK
gdMaxInitialization-Error	Maximum timing error that a node may have following integration.	0–11.7 μs	MAXINITIAL-IZATION-ERROR
gdMaxMicrotick	Maximum microtick length of all microticks configured within a cluster.	pdMicrotick (μs)	n.a.
gdMaxPropagation-Delay	Maximum propagation delay of a cluster.	<=2.5 μs	n.a.
gdMinislot	Duration of a minislot.	2–63 MT	MINISLOT
gdMinislotActionPoint-Offset	Number of macroticks the minislot action point is offset from the beginning of a minislot.	1–31 MT	MINISLOT-ACTION-POINT-OFFSET
gdMinPropagationDelay	Minimum propagation delay of a cluster.	<=gdMaxPr-opagation-Delay (μs)	n.a.
gdNIT	Duration of the network idle time.	2–805 MT	N-I-T
gdSampleClockPeriod	Sample clock period.	(0.0125, 0.025, 0.05 μs)	SAMPLE-CLOCK-PERIOD
gdStaticSlot	Duration of a static slot.	4–661 MT	STATIC-SLOT
gdSymbolWindow	Duration of the symbol window.	0–142 MT	SYMBOL-WINDOW

Table A2.2   (*continued*)

Name	Description	Range	FIBEX
gdTSSTransmitter	Number of bits in the transmission start sequence.	3–15 gdBit	T-S-S-TRANS-MITTER
gdWakeupSymbol-RxIdle	Number of bits used by the node to test the duration of the 'idle' portion of a received wakeup symbol. Duration is equal to (gdWakeupSymbolTxIdle – gdWakeupSymbolTxLow)/2 minus a safe part. (Collisions, clock differences and other effects can deform the Tx-wakeup pattern.)	14–59 gdBit	WAKE-UP-SYMBOL-RX-IDLE
gdWakeupSymbol-RxLow	Number of bits used by the node to test the LOW portion of a received wakeup symbol. This lower limit of zero bits has to be received to detect the LOW portion by the receiver. The duration is equal to gdWakeupSymbolTxLow minus a safe part. (Active stars, clock differences and other effects can deform the Tx-wakeup pattern.)	11–59 gdBit	WAKE-UP-SYMBOL-RX-LOW
gdWakeupSymbolRx-Window	The size of the window used to detect wakeups. Detection of a wakeup requires a low and idle period (from one WUS) and a low period (from another WUS) to be detected entirely within a window of this size. The duration is equal to gdWakeupSymbolTxIdle $+ 2 \times$ gdWakeupSymbolTxLow plus a safe part. (Clock differences and other effects can deform the Tx-wakeup pattern.)	76–301 gdBit	WAKE-UP-SYMBOL-RX-WINDOW
gdWakeupSymbol-TxIdle	Number of bits used by the node to transmit the 'idle' part of a wakeup symbol. The duration is equal to cdWakeupSymbolTxIdle.	45–180 gdBit	WAKE-UP-SYMBOL-TX-IDLE

(*continued overleaf*)

Table A2.2   (*continued*)

Name	Description	Range	FIBEX
gdWakeupSymbol-TxLow	Number of bits used by the node to transmit the LOW part of a wakeup symbol. The duration is equal to cdWakeupSymbolTxLow.	15–60 gdBit	WAKE-UP-SYMBOL-TX-LOW
gListenNoise	Upper limit for the startup listen timeout and wakeup listen timeout in the presence of noise. It is used as a multiplier of the node parameter pdListenTimeout.	2–16 times	Listen-Noise
gMacroPerCycle	Number of macroticks in a communication cycle.	10–16 000 MT	MACRO-PER-CYCLE
gMaxWithoutClock-CorrectionFatal	Threshold used for testing the vClockCorrectionFailed counter. Defines the number of consecutive even/odd cycle pairs with missing clock correction terms that will cause the protocol to transition from the POC:normal active or POC:normal passive state into the POC:halt state.	gMaxWithout-Clock-Correction-Passive-15 even/odd cycle pairs	MAX-WITHOUT-CLOCK-CORRECTION-FATAL
gMaxWithoutClock-CorrectionPassive	Threshold used for testing the vClockCorrectionFailed counter. Defines the number of consecutive even/odd cycle pairs with missing clock correction terms that will cause the protocol to transition from the POC:normal active state to the POC:normal passive state. Note that gMaxWithoutClockCorrectionPassive <= gMaxWithoutClockCorrectionFatal <= 15.	1–15 even/odd cycle pairs	MAX-WITHOUT-CLOCK-CORRECTION-PASSIVE
gNetwork-ManagementVector-Length	Length of the network management vector in a cluster.	0–12 bytes	NETWORK-MANAGEMENT-VECTOR-LENGTH
gNumberOfMinislots	Number of minislots in the dynamic segment.	0–7986	NUMBER-OF-MINISLOTS
gNumberOfStaticSlots	Number of static slots in the static segment.	2 – cStatic-SlotIDMax	NUMBER-OF-STATIC-SLOTS

Table A2.2   (*continued*)

Name	Description	Range	FIBEX
gOffsetCorrectionMax	Cluster global magnitude of the maximum necessary offset correction value.	0.15– 383.567 μs	OFFSET-CORRECTION-MAX
gOffsetCorrectionStart	Start of the offset correction phase within the NIT, expressed as the number of macroticks from the start of cycle.	9–15 999 MT	OFFSET-CORRECTION-START
gPayloadLengthStatic	Payload length of a static frame.	0–127 words	PAYLOAD-LENGTH-STATIC
gSyncNodeMax	Maximum number of nodes that may send frames with the sync frame indicator bit set to one.	2–15 nodes	SYNC-NODE-MAX

NIT = network idle time

Table A2.3   Parameters of the nodes

Name	Description	Range	FIBEX
pAllowHaltDueToClock	Boolean flag that controls the transition to the POC:halt state due to a clock synchronisation error. If set to true, the CC is allowed to transition to POC:halt. If set to false, the CC will not transition to the POC:halt state but will enter or remain in the POC:normal passive state (self healing would still be possible).	Boolean	ALLOW-HALT-DUE-TO-CLOCK
pAllowPassiveToActive	Number of consecutive even/odd cycle pairs that must have valid clock correction terms before the CC will be allowed to transition from the POC:normal passive state to the POC:normal active state. If set to zero, the CC is not allowed to transition from POC:normal passive to POC:normal active.	0–31 even/odd cycle pairs	ALLOW-PASSIVE-TO-ACTIVE

(*continued overleaf*)

Table A2.3   (*continued*)

Name	Description	Range	FIBEX
pChannels	Channels to which the node is connected.	(A, B, A and B)	CONNECTOR ID
pClusterDriftDamping	Local cluster drift damping factor used for rate correction.	0–20 μT	CLUSTER-DRIFT-DAMPING
pdAcceptedStartup-Range	Expanded range of measured clock deviation allowed for startup frames during integration.	0–1875 μT	ACCEPTED-STARTUP-RANGE
pDecodingCorrection	Value used by the receiver to calculate the difference between primary time reference point and secondary time reference point.	14–143 μT	DECODING-CORRECTION
pDelayCompensation[A], pDelay-Compensation[B]	Value used to compensate for reception delays on the indicated channel. This covers assumed propagation delay up to cPropagationDelayMax for microticks in the range of 0.0125–0.05 μs. In practice, the minimum of the propagation delays of all sync nodes should be applied.	0–200 μT	DELAY-COMPEN-SATION-A DELAY-COMPEN-SATION-B
pdListenTimeout	Value for the startup listen timeout and wakeup listen timeout. Although this is a node local parameter, the real time equivalent of this value should be the same for all nodes in the cluster.	1284–1 283 846 μT	LISTEN-TIMEOUT
pdMaxDrift	Maximum drift offset between two nodes that operate with unsynchronised clocks over one communication cycle.	2–1923 μT	MAX-DRIFT
pdMicrotick	Duration of a microtick.	pSamplesPer-Microtick × gdSample-ClockPeriod (μs)	MICROTICK

Table A2.3 (*continued*)

Name	Description	Range	FIBEX
pExternOffset-Correction	Number of microticks added or subtracted to the NIT to carry out a host-requested external offset correction.	0–7 µT	EXTERN-OFFSET-CORRECTION
pExternRateCorrection	Number of microticks added or subtracted to the cycle to carry out a host-requested external rate correction.	0–7 µT	EXTERN-RATE-CORRECTION
pKeySlotId	ID of the slot used to transmit the startup frame, sync frame or designated single slot frame.	1–1023 slots	KEY-SLOT-USAGE
pKeySlotUsedFor-Startup	Flag indicating whether the key slot is used to transmit a startup frame. If pKeySlotUsedForStartup is set to true then pKeySlotUsedForSync must also be set to true.	Boolean	STARTUP-SYNC
pKeySlotUsedForSync	Flag indicating whether the key slot is used to transmit a sync frame. If pKeySlotUsedForStartup is set to true then pKeySlotUsedForSync must also be set to true.	Boolean	SYNC
pLatestTx	Number of the last minislot in which a frame transmission can start in the dynamic segment.	0–7980 minislots	LATEST-TX
pMacroInitialOffset[A], pMacroInitial-Offset[B]	Integer number of macroticks between the static slot boundary and the following macrotick boundary of the secondary time reference point based on the nominal macrotick duration.	2–68 MT	MACRO-INITIAL-OFFSET-A MACRO-INITIAL-OFFSET-B
pMicroInitialOffset[A], pMicroInitial-Offset[B]	Number of microticks between the closest macrotick boundary described by pMacroInitialOffset[Ch] and the secondary time reference point. The parameter depends on pDelayCompensation[Ch] and therefore it has to be set independently for each channel.	0–239 µT	MICRO-INITIAL-OFFSET-A MICRO-INITIAL-OFFSET-B

(*continued overleaf*)

Table A2.3   (*continued*)

Name	Description	Range	FIBEX
pMicroPerCycle	Nominal number of microticks in the communication cycle of the local node. If nodes have different microtick durations this number will differ from node to node.	640–640 000 µT	MICRO-PER-CYCLE
pOffsetCorrectionOut	Magnitude of the maximum permissible offset correction value.	13–15 567 µT	OFFSET-CORRECTION-OUT
pPayloadLength-DynMax	Maximum payload length for dynamic frames.	0–127 words	MAX-DYNAMIC-PAYLOAD-LENGTH
pRateCorrectionOut	Magnitude of the maximum permissible rate correction value.	2–1923 µT	MAX-DRIFT
pSamplesPerMicrotick	Number of samples per microtick.	(1, 2, 4)	SAMPLES-PER-MICROTICK
pSingleSlotEnabled	Flag indicating whether or not the node shall enter single slot mode following startup.	Boolean	SINGLE-SLOT-ENABLED
pWakeupChannel	Channel used by the node to send a wakeup pattern.	(A, B)	WAKE-UP-CHANNEL
pWakeupPattern	Number of repetitions of the wakeup symbol that are combined to form a wakeup pattern when the node enters the POC:wakeup send state.	2–63 times	WAKE-UP-PATTERN

CC = communication controller

Bibliography

Böke, C. (2003) Automatic configuration of real time operating systems and real time communication systems for distributed embedded applications. PhD thesis. University of Paderborn.

Braberman, V.A. (2000) Modeling and checking real-time systems designs. PhD thesis. Departamento de Computacion, Universidad de Buenos Aires, Argentina.

Cena, G. and Valenzano, A. (2006) On the properties of the flexible time division multiple access technique. *IEEE Transactions on Industrial Informatics*, **2** (2).

Clément, C. (2004) Rapport bibliographique exécutifs temps réel, Mémoire de DEA systèmes Informatiques Répartis, Université Pierre et Marie Curie, Paris.

George, L. (1998) Ordonnancement en-ligne temps réel critique dans les systèmes distribués. PhD thesis in information technology. Université de Versailles St-Quentin.

George, L. (2005) Conditions de faisabilité pour l'ordonnancement temps réel préemptif et non préemptif. Proceedings of Ecole d'été Temps Réel.

Grenier, M. (2004) Nouvelles politiques d'ordonnancement obtenues par des techniques d'optimisation. Mémoire de DEA informatique, Université Henri Poincaré, Nancy.

Kopetz, H. (1997) *Real Time Systems-Design Principles for Distributed Embedded Applications*, Kluwer Academic Publishers.

Lawler, E.L. and Martel, C. (1981) Scheduling periodically occurring tasks on multiple processors. *Information Processing Letters*, **7**, 9–12.

Lehoczky, J.P. (1990) Fixed priority scheduling of periodic task sets with arbitrary deadlines. Proceedings of the 11th IEEE Real Time Systems Symposium.

Leung, J.Y.T. and Whitehead, J. (1982) On the complexity of fixed priority scheduling of periodic real time tasks. *Performance Evaluation*, **4** (2), 72.

Liu, C.L. and Layland, J.W. (1973) Scheduling algorithms for multiprogramming in a hard real-time environment. *Journal of the Association for Computing Machinery*, **20** (1), 46–61.

Paret, D. (2005) *Réseaux Multiplexés Pour Systèmes Embarqués: CAN, LIN, FlexRay, Safe-by-Wire*, edn Dunod, Collection Technique et Ingénierie.

Pop, T., Eles, P. and Peng, Z. (2004) Schedulability-driven communication synthesis for time triggered embedded systems. *Real Time Systems Journal*, (24), 297–325.

Pop, T., Pop, P., Eles, P. and Peng, Z. (2007) Bus access optimisation for FlexRay based distributed embedded systems. Proceedings in DATE.

Pop, T., Pop, P., Eles, P., Peng, Z. and Andrei, A. (2006) Timing analysis of the FlexRay communication protocol. Proceedings in ECRTS.

Sha, L. and Sathaye, S.S. (1995) Distributed System Design Using Generalized Rate Monotonic Theory. Technical report. Software Engineering Institute, Carnegie Mellon University, Pittsburgh, PA.

Tindell, K., Burns, A. and Wellings, A.J. (1991) Guaranteeing Hard Real Time End-to-end Communications Deadlines. Technical report RTRG/91/107, Department of Computer Science, University of York, England.

Tindell, K., Burns, A. and Wellings, A.J. (1995) Calculating Controller Area Network (CAN) Message Response Time. *Control Engineering Practice*, **3** (8), 1163–1169.

Tindell, K. and Clark, J. (1994) Holistic schedulability analysis for distributed hard real time systems. *Microprocessing and Microprogramming*, **40**, 117–134.

Wilwert, C. (2005) Influence des fautes transitoires et des performances temps réel sur la sureté des systèmes X by Wire. PhD thesis. Institut National Polytechnique de Lorraine.

Index

FlexRay and its Applications: Real Time Multiplexed Network, First Edition. Dominique Paret.
© 2012 John Wiley & Sons, Ltd. Published 2012 by John Wiley & Sons, Ltd.